I0409265

Overview

"One Giant Leap: the Untold Story of Apollo 11" is a comprehensive account of the historic Apollo 11 moon landing. From President Kennedy's declaration to the first step on the moon, this book provides a detailed overview of the entire mission. It explores the preparation and training of the astronauts, the crucial role of mission control, and the challenges faced along the way. The book also highlights the significant contributions of black women mathematicians, such as Katherine Johnson, Dorothy Vaughan, and Mary Jackson, and their impact on the success of the Apollo program. With a focus on the astronauts, including Neil Armstrong, Buzz Aldrin, and Michael Collins, the book delves into their experiences and reflections. Additionally, it examines the lasting legacy of Apollo 11, including scientific discoveries, technological advancements, and the inspiration it provided for future generations.

Table Of Contents

1 President Kennedy's Declaration

1.1 The Space Race Begins

The year was 1961, and the world was caught in the grip of the Cold War. The United States and the Soviet Union were locked in a fierce competition, not only for military supremacy but also for dominance in the realm of space exploration. It was against this backdrop that President John F. Kennedy made a historic declaration that would set the stage for one of the greatest achievements in human history.

On May 25, 1961, President Kennedy stood before a joint session of Congress and boldly proclaimed, "I believe that this nation should commit itself to achieving the goal, before this decade is out, of landing a man on the moon and returning him safely to the Earth." With those words, the race to the moon officially began.

1.1.1 The Soviet Union's Early Lead

At the time of President Kennedy's declaration, the Soviet Union had already taken a significant lead in the space race. They had successfully launched the first artificial satellite, Sputnik, into orbit in 1957, and just four years later, they sent Yuri Gagarin, the first human, into space. The United States, on the other hand, had experienced a series of setbacks and failures in its attempts to match the Soviet Union's achievements.

1.1.2 A National Imperative

President Kennedy recognized the need for the United States to regain its footing in the space race. He saw it not only as a matter of national pride but also as a strategic imperative. The space race was not just about exploring the unknown; it was about demonstrating technological superiority and the ability to project power beyond the Earth's atmosphere.

1.1.3 The Formation of NASA

To achieve this ambitious goal, President Kennedy knew that a new organization would be needed. In response to his call, the National Aeronautics and Space Administration (NASA) was established on July 29, 1958. NASA would be responsible for coordinating all of the United States' civilian space activities and leading the charge in the race to the moon.

1.1.4 The Mercury Program

In the early years of NASA, the focus was on developing the necessary technology and capabilities to send humans into space. This effort culminated in the Mercury program, which aimed to put an American astronaut into orbit around the Earth. On May 5, 1961, Alan Shepard became the first American to travel into space aboard the Freedom 7 spacecraft.

1.1.5 The Challenge of the Moon

While the Mercury program was a significant achievement, President Kennedy knew that reaching the moon would require a whole new level of technological prowess. The moon was over 238,000 miles away, and no human had ever ventured beyond Earth's orbit. The challenges were immense, but President Kennedy believed that with the right resources and determination, the United States could accomplish this feat.

1.1.6 A Race Against Time

President Kennedy's declaration set a clear deadline for the moon landing: before the end of the 1960s. This meant that NASA had less than a decade to develop the necessary spacecraft, launch vehicles, and support systems to make the mission a reality. The clock was ticking, and the pressure was on.

1.1.7 The Cold War Context

The space race was not just about scientific exploration; it was also a battle for ideological supremacy. The United States and the Soviet Union saw space as the next frontier for demonstrating their respective systems' superiority. Each successful mission was a propaganda victory, and each failure was a blow to national pride. The stakes were high, and the world was watching.

1.1.8 A New Era of Exploration

President Kennedy's declaration marked the beginning of a new era of exploration. It was a bold and audacious goal that captured the imagination of people around the world. The space race would push the boundaries of human knowledge and capabilities, and it would ultimately lead to one of the most remarkable achievements in history: the first manned moon landing.

As the United States embarked on this monumental journey, the world held its breath. The space race had begun, and the race to the moon was on. The next chapters of this book will delve into the preparation and training of the astronauts, the challenges faced by mission control, the journey to the moon, the landing itself, and the lasting legacy of Apollo 11. It is a story of human ingenuity, perseverance, and the triumph of the human spirit.

1.2 Kennedy's Vision for the Moon

In the early 1960s, the United States found itself in the midst of the Cold War with the Soviet Union. The two superpowers were engaged in a fierce competition to demonstrate their technological and ideological superiority. The Soviet Union had already achieved several significant milestones in space exploration, including launching the first satellite, Sputnik, and sending the first human, Yuri Gagarin, into orbit. These achievements sent shockwaves through the United States, and it became clear that the nation needed to respond with its own ambitious space program.

It was against this backdrop that President John F. Kennedy delivered a historic speech to a joint session of Congress on May 25, 1961. In his speech, Kennedy outlined his vision for the United States to land a man on the moon and return him safely to Earth before the end of the decade. This audacious goal, known as the Apollo program, would require an unprecedented level of technological innovation, scientific discovery, and human courage.

Kennedy's vision for the moon was not just about winning the space race; it was about asserting American leadership and demonstrating the country's commitment to exploration and discovery. He believed that the moon landing would serve as a symbol of American ingenuity and determination, showcasing the nation's capabilities to the world.

The President's declaration was met with both excitement and skepticism. Many questioned the feasibility of such an ambitious undertaking, while others saw it as a necessary response to the Soviet Union's space achievements. Regardless of the initial doubts, Kennedy's vision for the moon ignited a sense of national pride and unity. It galvanized the American people and inspired a generation of scientists, engineers, and astronauts.

To turn Kennedy's vision into reality, NASA, the National Aeronautics and Space Administration, had to overcome numerous challenges. The agency had to develop new technologies, design and build spacecraft capable of reaching the moon, and train a team of astronauts who would undertake this perilous journey.

Under the leadership of NASA Administrator James E. Webb, the agency set out to fulfill Kennedy's vision. Webb recognized the importance of the moon landing not only for scientific and technological advancement but also for its potential to inspire future generations. He rallied the nation's top scientists, engineers, and mathematicians to work tirelessly towards this common goal.

NASA's response to Kennedy's vision was multifaceted. The agency established the Apollo program, a series of missions that would culminate in the moon landing. The program involved the development of several spacecraft, including the command module, which would carry the astronauts to and from the moon, and the lunar module, which would land on the lunar surface.

To achieve the moon landing, NASA had to overcome significant technical challenges. The agency had to develop new propulsion systems, navigation techniques, and life support systems capable of sustaining astronauts in the harsh environment of space. It required the collaboration of thousands of scientists, engineers, and technicians who worked tirelessly to solve these complex problems.

One of the key aspects of Kennedy's vision was the involvement of human beings in the mission. Unlike the Soviet Union's unmanned lunar missions, the United States aimed to send astronauts to the moon and bring them back safely. This decision was not without risks, as the journey to the moon and back would expose the astronauts to extreme temperatures, radiation, and the unknown challenges of space travel.

Kennedy's vision for the moon also had a profound impact on the American public. It captured the imagination of people across the country and around the world. The moon landing became a shared goal, a symbol of human achievement that transcended national boundaries. It brought people together in a way that few other events in history have done.

As the Apollo program took shape, NASA began selecting the astronauts who would make the journey to the moon. These individuals would become the face of the mission, embodying the spirit of exploration and adventure. Among the most famous of these astronauts were Neil Armstrong, Buzz Aldrin, and Michael Collins, who would make up the crew of Apollo 11.

Kennedy's vision for the moon was not just about reaching a destination; it was about pushing the boundaries of human knowledge and capability. It was about proving that the impossible was possible. And on July 20, 1969, when Neil Armstrong took that historic first step onto the lunar surface and declared, "That's one small step for man, one giant leap for mankind," Kennedy's vision became a reality. The moon landing was a testament to the power of human ingenuity, determination, and the unwavering belief in the pursuit of the unknown.

1.3 NASA's Response

When President John F. Kennedy made his historic declaration on May 25, 1961, challenging the United States to send a man to the moon and return him safely to Earth before the end of the decade, it sparked a flurry of activity within NASA. The space agency, which had been established just three years earlier, now faced an immense task: to turn the President's vision into a reality.

NASA's response to President Kennedy's declaration was swift and determined. The agency recognized the significance of the challenge and understood that it would require a massive effort from all levels of the organization. The goal was not only to achieve a technological feat but also to demonstrate the United States' superiority in space exploration during the height of the Cold War.

To meet this challenge, NASA embarked on an ambitious program known as the Apollo Program. The program aimed to develop the necessary technology, spacecraft, and infrastructure to send astronauts to the moon and bring them back safely. It was a monumental undertaking that would push the boundaries of human knowledge and engineering capabilities.

Under the leadership of NASA Administrator James E. Webb, the agency set out to assemble a team of brilliant scientists, engineers, and technicians who would work tirelessly to make President Kennedy's vision a reality. The task was not easy, as it required the coordination of multiple departments within NASA and collaboration with various contractors and suppliers.

One of the first steps in NASA's response was to establish the Manned Spacecraft Center (MSC) in Houston, Texas. The MSC would serve as the hub for the Apollo Program, overseeing the development and testing of the spacecraft and training the astronauts who would eventually journey to the moon. The center would later be renamed the Johnson Space Center in honor of President Kennedy's commitment to space exploration.

NASA also recognized the need for a dedicated team of engineers and technicians who would be responsible for designing and building the spacecraft that would carry the astronauts to the moon. This led to the formation of the Apollo Spacecraft Program Office, which was tasked with overseeing the development of the Apollo Command/Service Module and the Lunar Module.

In addition to the technical challenges, NASA also had to address the logistical and operational aspects of the Apollo Program. This included establishing a network of tracking stations around the world to ensure continuous communication with the astronauts during their mission. It also involved developing new technologies for navigation, guidance, and control, as well as life support systems that would sustain the astronauts during their journey to the moon and back.

As NASA's response to President Kennedy's declaration gained momentum, the agency began to attract some of the brightest minds in the country. Engineers, scientists, and mathematicians flocked to NASA, eager to contribute their expertise to the Apollo Program. Among these talented individuals were a group of remarkable black women mathematicians, whose contributions would prove to be invaluable to the success of the Apollo 11 mission.

Katherine Johnson, Dorothy Vaughan, and Mary Jackson, known as the "Hidden Figures," were among the black women mathematicians who worked at NASA during the 1960s. These brilliant women played a crucial role in calculating the trajectories for the Apollo missions, including Apollo 11. Their calculations were essential for ensuring the accuracy of the spacecraft's trajectory and the successful navigation to the moon.

Despite facing discrimination and prejudice, these women persevered and made significant contributions to the Apollo Program. Their dedication and expertise helped pave the way for future generations of women and minorities in the field of mathematics and science.

As NASA's response to President Kennedy's declaration continued, the agency faced numerous challenges and setbacks. The tragic Apollo 1 fire in 1967 claimed the lives of astronauts Gus Grissom, Ed White, and Roger Chaffee and served as a stark reminder of the risks involved in space exploration. NASA learned valuable lessons from the tragedy and implemented rigorous safety measures to prevent similar incidents in the future.

Despite the setbacks, NASA remained committed to its goal of landing a man on the moon. The agency conducted several unmanned missions, including Apollo 8 and Apollo 10, to test the spacecraft and refine the procedures for the lunar landing. These missions provided valuable data and insights that would prove crucial for the success of Apollo 11.

In the years leading up to the historic Apollo 11 mission, NASA's response to President Kennedy's declaration was marked by determination, innovation, and collaboration. The agency overcame numerous challenges and setbacks to develop the technology, infrastructure, and expertise necessary to achieve the seemingly impossible. The stage was set for the next chapter in the Apollo Program: the preparation and training of the astronauts who would make history by taking the first steps on the moon.

1.4 The Apollo Program Takes Shape

As President Kennedy's vision for landing a man on the moon began to take shape, the Apollo program emerged as NASA's ambitious endeavor to achieve this monumental feat. The program aimed to develop the necessary spacecraft, technology, and infrastructure to make a lunar landing possible. With the declaration of this audacious goal, the United States embarked on a journey that would captivate the world and forever change the course of history.

1.4.1 Establishing the Apollo Program

Following President Kennedy's declaration on May 25, 1961, the Apollo program quickly gained momentum. NASA, under the leadership of Administrator James E. Webb, set out to assemble a team of brilliant minds and dedicated individuals to turn the dream of reaching the moon into a reality. The program's primary objective was to land astronauts on the lunar surface and return them safely to Earth.

1.4.2 The Mercury and Gemini Programs

Before the Apollo program could take shape, NASA had to lay the groundwork for human spaceflight. The Mercury program, which began in 1958, aimed to put an American astronaut into orbit around the Earth. This program successfully achieved its goal with the flight of John Glenn on February 20, 1962.

Building upon the success of Mercury, NASA launched the Gemini program in 1961. Gemini aimed to develop the necessary technology and techniques for longer-duration spaceflights, rendezvous, and docking maneuvers. Through a series of ten manned missions, Gemini paved the way for the Apollo program by testing crucial aspects of space travel and astronaut capabilities.

1.4.3 The Saturn V Rocket

Central to the success of the Apollo program was the development of the Saturn V rocket. This colossal rocket stood at a towering height of 363 feet and was capable of generating a staggering 7.5 million pounds of thrust. The Saturn V was a three-stage rocket, with each stage designed to propel the spacecraft further into space.

The first stage, known as the S-IC, was powered by five F-1 engines and provided the initial thrust needed to escape Earth's gravity. The second stage, called the S-II, was powered by five J-2 engines and propelled the spacecraft into Earth's orbit. Finally, the third stage, known as the S-IVB, was powered by a single J-2 engine and was responsible for the trans-lunar injection, sending the spacecraft on its trajectory towards the moon.

1.4.4 Lunar Module Development

Another critical component of the Apollo program was the development of the Lunar Module (LM). The LM was a two-stage spacecraft designed to transport astronauts from lunar orbit to the surface of the moon and back. It consisted of two sections: the descent stage, which provided the propulsion and landing gear, and the ascent stage, which housed the crew and the ascent engine.

The development of the LM presented numerous challenges, including the need for a lightweight yet robust design capable of withstanding the harsh lunar environment. NASA's engineers and contractors worked tirelessly to overcome these obstacles, resulting in the creation of a remarkable spacecraft that would enable astronauts to take their first steps on the lunar surface.

1.4.5 Mission Planning and Training

With the Apollo program taking shape, meticulous planning and rigorous training became paramount. NASA established a comprehensive training program for the astronauts, preparing them physically, mentally, and technically for the challenges they would face during their lunar mission.

The astronauts underwent extensive simulations and mock missions to familiarize themselves with the spacecraft and the procedures necessary for a successful lunar landing. These simulations allowed them to practice critical maneuvers, such as docking and rendezvous, as well as emergency procedures in case of system failures.

1.4.6 The Apollo Missions

The Apollo program consisted of a series of missions, each building upon the knowledge and experience gained from the previous ones. Apollo 1, unfortunately, ended in tragedy when a fire broke out during a pre-launch test, resulting in the loss of the entire crew. This devastating event served as a stark reminder of the risks involved in space exploration and prompted NASA to make significant safety improvements.

Apollo 8 marked a significant milestone in the program, as it became the first mission to orbit the moon. This mission provided valuable data and insights into the lunar environment, paving the way for subsequent missions that would ultimately lead to a lunar landing.

Apollo 10, often referred to as the "dress rehearsal" for the moon landing, saw the lunar module descend to within just miles of the lunar surface. This mission tested the LM's performance and the astronauts' ability to navigate and rendezvous with the command module in lunar orbit.

1.4.7 The Road to Apollo 11

With the successful completion of the earlier Apollo missions, the stage was set for the historic Apollo 11 mission. The crew, consisting of Commander Neil Armstrong, Lunar Module Pilot Buzz Aldrin, and Command Module Pilot Michael Collins, had been meticulously selected and trained for this momentous undertaking.

The Apollo 11 mission would be the culmination of years of planning, development, and training. The spacecraft, the Saturn V rocket, and the Lunar Module had all been thoroughly tested and refined to ensure the crew's safety and the mission's success. The world held its breath as the countdown to launch began, eagerly anticipating the moment when humanity would take its first steps on another celestial body.

The Apollo program had taken shape, and the world watched in awe as the United States prepared to achieve what was once thought impossible. The stage was set for the most extraordinary journey ever undertaken by humankind, as Apollo 11 embarked on its mission to make history and forever change our understanding of the universe.

2 Preparation and Training

2.1 Selecting the Astronauts

The success of the Apollo 11 mission relied heavily on the selection of the right individuals to embark on this historic journey. The process of choosing the astronauts was meticulous and rigorous, ensuring that only the most qualified and capable individuals would have the opportunity to venture into space.

NASA's selection criteria for astronauts were stringent. They sought individuals who possessed exceptional physical fitness, intelligence, and mental fortitude. The astronauts needed to be able to withstand the physical and psychological demands of space travel, as well as possess the ability to work effectively as a team in high-pressure situations.

The initial pool of candidates was vast, with hundreds of highly qualified individuals vying for the opportunity to be part of the Apollo program. NASA's selection committee, led by Deke Slayton, carefully reviewed each candidate's qualifications, including their education, flight experience, and military service. They also considered personal attributes such as adaptability, problem-solving skills, and leadership potential.

Ultimately, only a select few were chosen to become astronauts. The Apollo 11 crew consisted of three remarkable individuals: Neil Armstrong, Buzz Aldrin, and Michael Collins. Each astronaut brought unique skills and experiences to the mission, making them an exceptional team.

Neil Armstrong, the mission commander, was a seasoned test pilot with extensive experience in aeronautics. He had flown numerous high-speed aircraft and had a reputation for his calm demeanor and exceptional piloting skills. Armstrong's ability to remain composed under pressure would prove invaluable during the mission's critical moments.

Buzz Aldrin, the lunar module pilot, was a brilliant engineer and a decorated fighter pilot. His expertise in orbital mechanics and his experience as a combat pilot made him an ideal candidate for the Apollo program. Aldrin's analytical mind and attention to detail would play a crucial role in the success of the mission.

Michael Collins, the command module pilot, was an accomplished astronaut with a background in both military aviation and spaceflight. Collins had previously participated in the Gemini 10 mission, where he demonstrated his exceptional piloting skills and ability to perform complex tasks in space. His role in Apollo 11 was to remain in orbit around the moon while Armstrong and Aldrin descended to the lunar surface.

The selection of these three astronauts was not only based on their individual qualifications but also on their ability to work together as a cohesive unit. The Apollo 11 crew underwent extensive training and simulations to ensure they could effectively communicate and collaborate during the mission. Their ability to trust and rely on one another was paramount to the success of the mission.

While the focus of the Apollo 11 mission was on the astronauts, it is important to acknowledge the critical contributions of the individuals in mission control. These dedicated professionals worked tirelessly behind the scenes, monitoring the spacecraft's systems, providing guidance to the astronauts, and making split-second decisions to ensure the mission's success.

The team in mission control was led by flight directors such as Gene Kranz, who became synonymous with the Apollo program. Kranz and his team were responsible for overseeing every aspect of the mission, from launch to landing. Their expertise and quick thinking were instrumental in resolving any issues that arose during the mission.

In addition to the astronauts and mission control, it is essential to recognize the significant contributions of black women mathematicians to the Apollo program. Katherine Johnson, Dorothy Vaughan, and Mary Jackson, known as the "Hidden Figures," played pivotal roles in calculating the trajectories and orbits necessary for the success of the Apollo missions.

These brilliant mathematicians overcame numerous challenges and barriers to excel in their fields. Their calculations were crucial in ensuring the accuracy of the spacecraft's trajectory and the timing of critical maneuvers. Their contributions were invaluable and demonstrated the importance of diversity and inclusion in scientific endeavors.

The selection of the astronauts for the Apollo 11 mission was a meticulous process that resulted in the formation of a remarkable team. Neil Armstrong, Buzz Aldrin, and Michael Collins, along with the support of the mission control team and the contributions of the "Hidden Figures," would go on to make history as they embarked on the journey to the moon. Their selection was a testament to their exceptional skills, dedication, and unwavering commitment to the success of the Apollo program.

2.2 Training for the Unknown

Once the team of astronauts had been selected for the Apollo 11 mission, they embarked on a rigorous training program to prepare themselves for the unknown challenges they would face during their journey to the moon. The training was designed to simulate the conditions they would encounter in space and to ensure that they were fully equipped to handle any situation that may arise.

The training program began with a focus on physical fitness. The astronauts underwent intense physical conditioning to prepare their bodies for the physical demands of space travel. They participated in a variety of exercises, including running, swimming, and weightlifting, to build strength and endurance. They also underwent extensive medical examinations to ensure that they were in optimal health for the mission.

In addition to physical training, the astronauts also underwent extensive classroom instruction. They studied a wide range of subjects, including celestial navigation, spacecraft systems, and lunar geology. They learned how to operate the various instruments and equipment they would be using during the mission, as well as how to troubleshoot any potential issues that may arise.

One of the most critical aspects of the training program was the simulation and mock mission exercises. These exercises were designed to replicate the conditions the astronauts would experience during their mission and to test their ability to respond to various scenarios. They practiced everything from launch procedures to docking maneuvers to lunar surface operations.

The astronauts spent countless hours in simulators, which replicated the spacecraft and its controls. They practiced flying the spacecraft, docking with the lunar module, and performing extravehicular activities on the lunar surface. These simulations allowed them to become familiar with the spacecraft's systems and to develop the muscle memory necessary to operate them effectively.

In addition to the simulators, the astronauts also participated in mock missions. These missions were conducted in remote locations that closely resembled the lunar surface. The astronauts would spend several days living and working in these simulated lunar environments, conducting experiments and performing tasks that they would later perform on the actual moon.

The training program also included survival training. The astronauts learned how to survive in various environments, including the desert and the ocean, in the event of an emergency landing. They were taught how to find food and water, build shelters, and signal for rescue. This training was essential in case the astronauts were forced to abort their mission and make an emergency landing on Earth.

Throughout their training, the astronauts were constantly evaluated and assessed. Their performance was closely monitored, and any weaknesses or areas for improvement were identified and addressed. The training program was designed to push the astronauts to their limits and to ensure that they were fully prepared for the challenges they would face during their mission.

As the launch date approached, the intensity of the training program increased. The astronauts underwent final simulations and mock missions to fine-tune their skills and to ensure that they were ready for the mission. They also participated in extensive briefings with mission control and support staff to ensure that everyone was on the same page and fully prepared for the mission.

The training program for the Apollo 11 mission was a testament to the dedication and commitment of the astronauts and the entire NASA team. It was a grueling and demanding process, but it was necessary to ensure the success of the mission. The training program prepared the astronauts for the unknown challenges they would face during their journey to the moon and gave them the confidence and skills they needed to make history.

2.3 Simulations and Mock Missions

Before embarking on the historic journey to the moon, the astronauts of Apollo 11 underwent extensive simulations and mock missions to prepare themselves for the challenges they would face in space. These simulations played a crucial role in ensuring the success of the mission and the safety of the crew.

2.3.1 Simulating Space Conditions

To recreate the conditions of space, NASA built specialized simulators that replicated the environment inside the spacecraft. These simulators, such as the Lunar Module Simulator and the Command Module Simulator, allowed the astronauts to familiarize themselves with the controls, systems, and procedures they would encounter during the mission.

Inside these simulators, the astronauts practiced various scenarios, including launch procedures, docking maneuvers, and emergency situations. They rehearsed every step of the mission repeatedly, ensuring that they were well-prepared for any eventuality. These simulations helped the astronauts develop muscle memory and quick reflexes, enabling them to respond effectively in the challenging and unpredictable conditions of space.

2.3.2 Mock Missions

In addition to the simulations, the astronauts also participated in mock missions that closely resembled the actual lunar mission. These mock missions, known as "dress rehearsals," aimed to replicate the entire mission timeline, from launch to lunar landing and return.

During these mock missions, the astronauts followed the same procedures and timelines as they would during the actual mission. They practiced every aspect of the mission, including communication with mission control, spacecraft maneuvers, and lunar surface operations. These dress rehearsals allowed the astronauts to identify and address any potential issues or challenges that could arise during the mission.

The mock missions also involved collaboration with mission control and the support staff. This collaboration was crucial in ensuring effective communication and coordination between the astronauts and the ground team. It allowed both the astronauts and mission control to refine their procedures and protocols, ensuring a seamless flow of information and decision-making during the actual mission.

2.3.3 Training for Contingencies

One of the primary objectives of the simulations and mock missions was to prepare the astronauts for contingencies and emergencies that could arise during the mission. The astronauts trained for a wide range of scenarios, including system failures, navigation errors, and medical emergencies.

In these training sessions, the astronauts learned how to troubleshoot and resolve technical issues that could potentially jeopardize the mission. They practiced emergency procedures, such as aborting a landing or executing a safe return to Earth in case of critical failures. These simulations and mock missions helped the astronauts develop the skills and confidence needed to handle unexpected situations and make split-second decisions in high-pressure environments.

2.3.4 Psychological Preparation

Apart from the technical aspects, the simulations and mock missions also played a crucial role in psychologically preparing the astronauts for the challenges of space travel. Spending extended periods in isolation, confined spaces, and weightlessness can have profound psychological effects on individuals.

The simulations allowed the astronauts to experience the physical and psychological sensations of space travel, helping them adapt to the unique environment they would encounter during the mission. They learned techniques to manage stress, maintain focus, and cope with the isolation and confinement of the spacecraft.

Furthermore, the simulations and mock missions fostered a strong sense of camaraderie and teamwork among the astronauts. They worked closely with their fellow crew members, relying on each other for support and cooperation. This teamwork was essential for the success of the mission, as it ensured effective communication and collaboration during critical moments.

In conclusion, the simulations and mock missions undertaken by the Apollo 11 astronauts were instrumental in preparing them for the challenges they would face during their historic journey to the moon. These training exercises allowed the astronauts to familiarize themselves with the spacecraft, practice mission procedures, and develop the skills and confidence needed to overcome contingencies. The simulations not only prepared them technically but also psychologically, ensuring that they were ready to face the unknown and make history on that fateful day in July 1969.

2.4 Preparing for Liftoff

As the Apollo 11 mission drew closer, the anticipation and excitement within NASA and the astronaut team reached its peak. The meticulous preparation and training that had taken place over the years were now being put to the ultimate test. The astronauts, Neil Armstrong, Buzz Aldrin, and Michael Collins, along with the dedicated team at mission control, were all focused on one goal: a successful liftoff and journey to the moon.

2.4.1 Final Checks and Preparations

In the days leading up to the launch, the spacecraft underwent a series of final checks and preparations. Every system and component were meticulously inspected to ensure they were in perfect working order. The engineers and technicians worked tirelessly, leaving no stone unturned in their quest for perfection.

The astronauts themselves were involved in the final checks, going through extensive checklists to ensure they were familiar with every aspect of the spacecraft. They reviewed emergency procedures, practiced donning their spacesuits, and familiarized themselves with the layout of the command module and lunar module.

2.4.2 Suiting Up for the Journey

On the day of the launch, the astronauts arrived at the Kennedy Space Center, ready to suit up for their historic journey. They donned their iconic white spacesuits, specially designed to protect them from the harsh conditions of space. The suits were not only airtight but also equipped with cooling systems to regulate their body temperature.

The process of suiting up was meticulous and time-consuming. Each astronaut had a team of technicians assisting them, ensuring that every component of the suit was properly fitted and functioning. The helmets were secured, gloves were checked for dexterity, and the astronauts were connected to the life support systems that would sustain them during their time in space.

2.4.3 Preparing the Command Module

While the astronauts were getting suited up, the command module, named Columbia, was being prepared for liftoff. The spacecraft had undergone extensive testing and was now ready to carry the astronauts to the moon and back. The technicians made final checks on the communication systems, navigation instruments, and life support systems.

Inside the command module, the astronauts' seats were meticulously adjusted to ensure their comfort and safety during the intense acceleration of liftoff. The control panels were checked, and the astronauts familiarized themselves with the layout of the switches and buttons they would need to operate during the mission.

2.4.4 Countdown and Liftoff

As the countdown to liftoff began, the tension in the air was palpable. The astronauts, now fully suited up and seated inside the command module, went through their final checks. The mission control team closely monitored every system, ensuring that all parameters were within the acceptable range.

At T-minus 9 minutes and counting, the engines of the Saturn V rocket ignited, filling the air with a deafening roar. The ground shook as the massive rocket slowly lifted off the launchpad, defying gravity. The astronauts felt the intense acceleration as they were propelled into space, leaving Earth behind.

2.4.5 Mission Control's Role

While the astronauts were on their way to the moon, the team at mission control played a critical role in monitoring the spacecraft's systems and providing support. Flight directors, such as Gene Kranz, led the teams of engineers and support staff, ensuring that everything was running smoothly.

Mission control was a hive of activity, with multiple screens displaying vital information about the spacecraft's trajectory, fuel consumption, and life support systems. The flight controllers communicated with the astronauts, relaying instructions and updates. They worked tirelessly, analyzing data and making critical decisions to ensure the success of the mission.

2.4.6 Overcoming Challenges

Throughout the preparation and liftoff, the Apollo 11 team faced numerous challenges. From technical glitches to unforeseen issues, each obstacle had to be overcome swiftly and decisively. The astronauts and mission control worked together seamlessly, relying on their extensive training and expertise to troubleshoot and find solutions.

One of the most notable challenges during the liftoff was the communication blackout period. As the spacecraft ascended, it passed through a zone where the radio signals were blocked by the Earth's atmosphere. This meant that mission control temporarily lost contact with the astronauts. However, the team had prepared for this eventuality and had established contingency plans to ensure a smooth reestablishment of communication.

2.4.7 The Final Moments before Leaving Earth

As the spacecraft hurtled towards space, the astronauts experienced a mix of excitement, anticipation, and a touch of apprehension. They were leaving the safety of Earth's atmosphere and embarking on a journey into the unknown. The realization of the magnitude of their mission sank in, and they focused on the tasks ahead.

Inside the command module, the astronauts prepared for the next phase of their journey: the docking with the lunar module and the long voyage to the moon. They meticulously followed the procedures, double-checking every step to ensure a seamless transition. The teamwork and dedication of the entire Apollo 11 team were evident as they worked together to make history.

The liftoff of Apollo 11 was not just a momentous event for the astronauts and mission control; it was a defining moment for humanity. The successful liftoff marked the beginning of a journey that would take mankind to the moon and forever change our understanding of the universe.

3 Mission Control

3.1 The Unsung Heroes

While the astronauts of Apollo 11 often take center stage in the retelling of the historic moon landing, there were countless individuals behind the scenes who played crucial roles in the success of the mission. These unsung heroes, working tirelessly in mission control, were responsible for ensuring the safety and success of the astronauts as they embarked on their journey to the moon.

3.1.1 The Mission Control Team

At the heart of the Apollo 11 mission was the dedicated team of flight controllers and support staff in mission control. Led by flight director Gene Kranz, this team was responsible for monitoring every aspect of the mission, from liftoff to landing and everything in between. They were the ones who made split-second decisions, solved problems, and guided the astronauts through the challenges they faced.

The flight controllers, each specializing in a specific area such as guidance, navigation, or communications, worked in shifts around the clock to ensure continuous support for the mission. They were the ones who communicated with the astronauts, provided them with critical information, and helped them troubleshoot any issues that arose.

3.1.2 The Unsung Heroes

Beyond the flight controllers, there were numerous other individuals who played vital roles in the success of Apollo 11. Engineers, technicians, scientists, and support staff all worked together to make the mission possible.

The engineers and technicians were responsible for designing, building, and testing the spacecraft and its components. They meticulously crafted each part, ensuring that it met the rigorous standards required for space travel. Their attention to detail and commitment to quality control were essential in ensuring the safety of the astronauts.

Scientists and researchers played a crucial role in gathering and analyzing data throughout the mission. They monitored the spacecraft's systems, studied the lunar environment, and collected valuable scientific data that would contribute to our understanding of the moon and the universe.

Support staff, including administrative personnel, logistics experts, and medical professionals, provided the necessary infrastructure and support to keep the mission running smoothly. They coordinated schedules, managed supplies, and ensured the astronauts' well-being before, during, and after the mission.

3.1.3 The Role of Black Women Mathematicians

In addition to the dedicated individuals in mission control, the Apollo 11 mission also owes a debt of gratitude to a group of remarkable black women mathematicians. Katherine Johnson, Dorothy Vaughan, and Mary Jackson, known as the "Hidden Figures," made significant contributions to the success of the Apollo program.

Katherine Johnson, a brilliant mathematician, calculated the trajectories for the Apollo missions, including Apollo 11. Her precise calculations were crucial in ensuring that the spacecraft reached its intended destination with pinpoint accuracy.

Dorothy Vaughan, a skilled mathematician and computer programmer, led a team of African American women who worked as "human computers" at NASA. She played a vital role in transitioning the agency from human computers to electronic computers, ensuring the smooth operation of complex calculations required for the Apollo missions.

Mary Jackson, an aerospace engineer, worked on the development of the Apollo spacecraft. Her expertise and determination helped overcome barriers and paved the way for future generations of women and minorities in the field of engineering.

These remarkable women faced numerous challenges and overcame systemic barriers to make their mark on the Apollo program. Their contributions were instrumental in the success of Apollo 11 and the subsequent moon missions.

3.1.4 The Spirit of Collaboration

One of the remarkable aspects of the Apollo 11 mission was the spirit of collaboration that permeated every aspect of the project. People from diverse backgrounds and disciplines came together, united by a common goal: to achieve the seemingly impossible.

The collaboration extended beyond the walls of mission control and NASA. International partners, including scientists and engineers from around the world, contributed their expertise and resources to the mission. This global cooperation not only enhanced the scientific value of the mission but also fostered a sense of unity and shared purpose among nations.

The success of Apollo 11 was a testament to the dedication, ingenuity, and teamwork of all those involved. From the flight controllers in mission control to the engineers, technicians, and support staff, and the invaluable contributions of the black women mathematicians, each played a vital role in making the moon landing a reality. Their collective efforts and unwavering commitment to the mission ensured that Neil Armstrong and Buzz Aldrin could take that historic first step on the lunar surface, forever changing the course of human history.

3.2 Flight Directors and Support Staff

The success of the Apollo 11 mission relied not only on the bravery and skill of the astronauts, but also on the dedicated team of flight directors and support staff working tirelessly behind the scenes at NASA's Mission Control Center. These individuals played a crucial role in ensuring the safe and successful execution of the mission.

3.2.1 The Flight Directors

At the helm of Mission Control were the flight directors, responsible for making critical decisions and guiding the mission from liftoff to splashdown. Four flight directors were assigned to the Apollo 11 mission, each leading a specific phase of the mission.

The lead flight director for the mission was Gene Kranz, a seasoned veteran who had previously served as the flight director for the Gemini and Apollo missions. Kranz was known for his calm and decisive leadership style, earning him the respect and admiration of his team. He was responsible for overseeing the entire mission, from launch to landing.

Joining Kranz in the Mission Control Center were three other flight directors: Glynn Lunney, Gerry Griffin, and Milt Windler. Lunney was in charge of the ascent phase of the mission, overseeing the crucial moment when the lunar module would separate from the command module and begin its journey to the moon. Griffin was responsible for the lunar landing phase, ensuring a safe touchdown on the lunar surface. Windler, on the other hand, oversaw the return phase of the mission, including the reentry and splashdown.

3.2.2 The Support Staff

Behind the flight directors were a dedicated team of support staff, including flight controllers, engineers, and technicians. These individuals worked tirelessly to monitor the spacecraft's systems, communicate with the astronauts, and troubleshoot any issues that arose during the mission.

The flight controllers were responsible for monitoring various systems on the spacecraft, such as propulsion, navigation, and life support. They worked in shifts, ensuring round-the-clock coverage and constant vigilance. These controllers were highly trained and had to make split-second decisions in high-pressure situations.

In addition to the flight controllers, there were also engineers and technicians who played critical roles in the success of the mission. These individuals were responsible for designing, building, and testing the spacecraft and its components. They worked closely with the flight controllers to ensure that the systems were functioning properly and that any issues were addressed promptly.

3.2.3 Collaboration and Communication

Effective collaboration and communication were essential for the smooth operation of Mission Control. The flight directors and support staff worked closely together, sharing information and making decisions collectively. They relied on a complex network of communication systems to stay connected with the astronauts and each other.

One of the key communication systems used during the mission was the Apollo Unified S-Band System (AUS). This system allowed for voice and data communication between the spacecraft and Mission Control. It enabled the flight directors and support staff to relay instructions, receive status updates, and troubleshoot any issues that arose during the mission.

In addition to the AUS, Mission Control also utilized a network of tracking stations located around the world. These stations, operated by various countries, provided continuous tracking and communication with the spacecraft as it orbited the Earth and journeyed to the moon. This global network ensured that the flight directors and support staff had real-time information and could make informed decisions.

3.2.4 Problem Solving and Crisis Management

Throughout the Apollo 11 mission, the flight directors and support staff faced numerous challenges and crises. From minor technical glitches to potentially life-threatening situations, they had to think quickly and work together to find solutions.

One of the most notable crises occurred during the lunar landing phase. As the lunar module descended towards the moon's surface, the astronauts realized that they were heading towards a field of boulders. With limited fuel and time, the flight directors and support staff had to make a critical decision. They instructed the astronauts to abort the landing and ascend back to the command module. This decision, made under immense pressure, ultimately saved the lives of the astronauts.

Throughout the mission, the flight directors and support staff demonstrated their ability to remain calm and focused in the face of adversity. Their expertise, teamwork, and problem-solving skills were instrumental in overcoming the challenges that arose during the mission.

In conclusion, the flight directors and support staff of the Apollo 11 mission played a vital role in the success of the mission. Their dedication, expertise, and ability to work under pressure ensured the safe and successful execution of the first moon landing. Without their contributions, the historic achievement of Neil Armstrong, Buzz Aldrin, and Michael Collins would not have been possible.

3.3 Monitoring Systems and Communications

The success of the Apollo 11 mission relied heavily on the monitoring systems and communications that were established to ensure the safety and success of the astronauts. These systems played a crucial role in providing real-time information and facilitating communication between the spacecraft and mission control.

3.3.1 Tracking and Telemetry

One of the key components of the monitoring systems was the tracking and telemetry network. This network consisted of a series of ground-based tracking stations strategically located around the world. These stations were responsible for receiving and transmitting signals to and from the spacecraft.

The tracking stations utilized a combination of radar and radio frequency systems to track the position and trajectory of the spacecraft. This information was then relayed to mission control, allowing the flight controllers to monitor the progress of the mission and make any necessary adjustments.

In addition to tracking the spacecraft's location, the telemetry system provided vital data on various aspects of the mission. This included information on the spacecraft's systems, such as its propulsion, electrical, and life support systems. The telemetry data was continuously monitored by the flight controllers to ensure that everything was functioning as expected.

3.3.2 Communication Systems

Effective communication between the astronauts and mission control was essential for the success of the mission. To facilitate this, the Apollo spacecraft was equipped with a sophisticated communication system that allowed for both voice and data transmission.

The primary communication link between the spacecraft and mission control was established through the use of high-frequency radio waves. The spacecraft's antennas were designed to transmit and receive signals over long distances, enabling communication even when the spacecraft was on the far side of the Moon.

To ensure uninterrupted communication, multiple tracking stations were used to maintain contact with the spacecraft as it orbited the Earth or traveled to the Moon. These stations were strategically positioned to provide overlapping coverage, ensuring that there was always a line of sight between the spacecraft and at least one tracking station.

In addition to voice communication, the astronauts also relied on data transmission for various purposes. This included sending telemetry data to mission control, as well as receiving instructions and updates on mission objectives. The data transmission capabilities of the communication system played a crucial role in facilitating real-time decision-making and problem-solving during the mission.

3.3.3 Mission Control Center

The monitoring systems and communication networks were all coordinated and managed from the Mission Control Center located at the NASA Johnson Space Center in Houston, Texas. The Mission Control Center served as the nerve center for the Apollo 11 mission, with flight controllers working around the clock to monitor the spacecraft and provide guidance to the astronauts.

The Mission Control Center was divided into several key areas, each responsible for specific aspects of the mission. The Flight Operations Directorate oversaw the overall mission planning and execution, while the Flight Dynamics Officer monitored the spacecraft's trajectory and made course corrections as necessary.

The Communications and Data Handling Division was responsible for managing the communication systems and ensuring that the astronauts and mission control could effectively communicate with each other. This division also handled the transmission and reception of telemetry data, which was crucial for monitoring the health and status of the spacecraft.

The Flight Controllers, who were experts in their respective fields, worked in shifts to provide continuous monitoring and support. They were responsible for analyzing the data received from the spacecraft, making critical decisions, and relaying instructions to the astronauts.

3.3.4 Real-Time Decision Making

The monitoring systems and communication networks played a vital role in enabling real-time decision-making during the Apollo 11 mission. The flight controllers in mission control relied on the data received from the spacecraft to assess the mission's progress and make necessary adjustments.

In the event of any anomalies or emergencies, the flight controllers had to quickly analyze the data and determine the best course of action. This required a high level of expertise and the ability to think on their feet. The flight controllers worked closely with the astronauts, providing them with guidance and instructions to resolve any issues that arose.

The monitoring systems and communication networks also allowed for effective coordination between the astronauts and mission control during critical mission phases, such as lunar descent and ascent. The flight controllers provided the astronauts with real-time updates and guidance, ensuring that they were able to navigate the spacecraft safely.

Overall, the monitoring systems and communication networks established for the Apollo 11 mission were instrumental in ensuring the success of the mission. They provided the necessary data and communication capabilities that allowed mission control to monitor the spacecraft's progress, make informed decisions, and support the astronauts throughout their historic journey to the Moon.

3.4 Crises and Problem Solving

The Apollo 11 mission was an extraordinary feat of human achievement, but it was not without its fair share of crises and challenges. As the astronauts embarked on their historic journey to the moon, they encountered numerous obstacles that required quick thinking, problem-solving, and the expertise of the mission control team.

3.4.1 Navigating the Unknown

The Apollo 11 mission was a pioneering endeavor, venturing into uncharted territory. As the spacecraft hurtled through space, the crew and mission control faced a multitude of potential crises. They had to be prepared for any eventuality, from technical malfunctions to unforeseen emergencies.

3.4.2 The Oxygen Tank Incident

One of the most critical crises during the Apollo 11 mission occurred on the way to the moon. On April 13, 1970, an oxygen tank in the service module of Apollo 13 exploded, causing a catastrophic failure. The explosion not only jeopardized the lives of the astronauts but also threatened the entire mission.

3.4.3 Problem-Solving Under Pressure

The explosion of the oxygen tank presented a life-threatening situation for the crew of Apollo 13. The astronauts, Jim Lovell, Fred Haise, and Jack Swigert, along with the mission control team, had to work together to find solutions to the numerous challenges they faced.

3.4.4 Assessing the Damage

The first step in problem-solving was to assess the extent of the damage caused by the explosion. Mission control worked closely with the astronauts to gather information about the condition of the spacecraft and the available resources. They had to determine the best course of action to ensure the safe return of the crew.

3.4.5 Improvisation and Innovation

With limited resources and a damaged spacecraft, the crew and mission control had to think outside the box. They improvised solutions using the materials and equipment available on board. For example, they used duct tape, plastic bags, and cardboard to create makeshift filters for the carbon dioxide removal system.

3.4.6 Collaborative Problem-Solving

The crisis on Apollo 13 required a collaborative effort between the astronauts and mission control. The crew relayed information about the situation on board, while the experts on the ground analyzed the data and provided guidance. This close collaboration was crucial in finding solutions to the complex problems they faced.

3.4.7 Mission Control's Role

Mission control played a pivotal role in guiding the crew through the crisis. Flight directors, including Gene Kranz, Glynn Lunney, and Gerry Griffin, led the team of experts who worked tirelessly to develop strategies and procedures to overcome the challenges. Their calm and focused leadership provided the astronauts with the support they needed.

3.4.8 The Power of Simulations

The extensive training and simulations undertaken by the astronauts and mission control proved invaluable during the crisis. The simulations had prepared them for various scenarios, allowing them to draw on their knowledge and experience to find solutions. The ability to think quickly and adapt to changing circumstances was crucial in resolving the crisis.

3.4.9 The Successful Return

Through the collective efforts of the crew and mission control, Apollo 13 safely returned to Earth on April 17, 1970. The crisis had tested the limits of human ingenuity and problem-solving skills, but it also showcased the resilience and determination of the individuals involved.

3.4.10 Lessons Learned

The Apollo 13 crisis served as a valuable learning experience for NASA and the entire space program. It highlighted the importance of thorough preparation, effective communication, and the ability to adapt in the face of adversity. The lessons learned from this crisis would go on to shape future missions and improve the safety and success of space exploration.

3.4.11 The Spirit of Problem-Solving

The crises faced during the Apollo 11 mission and, in particular, the Apollo 13 incident, demonstrated the indomitable spirit of problem-solving that defined the space program. The ability to think critically, work collaboratively, and remain calm under pressure were essential qualities that enabled the astronauts and mission control to overcome seemingly insurmountable challenges.

3.4.12 A Testament to Human Ingenuity

The crises and problem-solving during the Apollo 11 mission showcased the remarkable capabilities of human ingenuity. The ability to overcome adversity and find innovative solutions in the face of danger is a testament to the indomitable spirit of exploration and the unwavering commitment to pushing the boundaries of human knowledge.

In the next section, we will delve into the final stages of the Apollo 11 mission, as the countdown to the moon landing begins and the astronauts prepare for the historic moment of stepping onto the lunar surface.

3.5 The Countdown Begins

As the day of the Apollo 11 mission approached, the excitement and anticipation grew exponentially. The world held its breath, waiting for the historic moment when humans would set foot on the moon for the very first time. But before that could happen, there were crucial steps that needed to be taken, and the countdown to launch began.

3.5.1 Preparing for Launch

In the days leading up to the launch, the astronauts and the mission control team worked tirelessly to ensure that everything was in place for a successful mission. The spacecraft underwent final checks and preparations, and the crew went through extensive training and simulations to familiarize themselves with every aspect of the mission.

The countdown to launch was a meticulously planned and executed process. It involved a series of checks and tests to ensure that all systems were functioning properly and that the spacecraft was ready for the journey to the moon. The countdown began at T-56 hours, and each milestone was carefully monitored and verified.

3.5.2 The Role of Mission Control

Mission Control played a critical role in the countdown process. Located at the Manned Spacecraft Center (now known as the Johnson Space Center) in Houston, Texas, Mission Control was the nerve center of the Apollo 11 mission. It was here that flight directors and support staff monitored the spacecraft's systems, communicated with the astronauts, and made critical decisions throughout the mission.

During the countdown, Mission Control closely monitored the spacecraft's systems to ensure that everything was functioning as expected. They checked the communication systems, life support systems, and the guidance and navigation systems. Any anomalies or issues that arose were quickly addressed and resolved to ensure the safety of the crew.

3.5.3 The Final Hours

As the countdown reached its final hours, the tension in Mission Control and among the astronauts was palpable. Every minute detail was scrutinized, and the crew went through their final preparations. The astronauts donned their spacesuits, checked their equipment, and mentally prepared themselves for the journey ahead.

At T-9 minutes and counting, the launch vehicle's guidance system was activated, and the final checks were conducted. The astronauts were strapped into their seats, and the spacecraft was powered up for launch. The countdown continued, and the tension in Mission Control reached its peak.

3.5.4 Liftoff

Finally, the moment arrived. At T-3 seconds, the engines ignited, and the Saturn V rocket roared to life. The ground shook, and the immense power of the rocket was unleashed. The spacecraft lifted off the launch pad, defying gravity and embarking on its historic journey to the moon.

The atmosphere in Mission Control was electric as the flight directors and support staff closely monitored the spacecraft's ascent. They tracked its trajectory, ensuring that it was on the correct path and that all systems were functioning properly. The countdown had culminated in a successful liftoff, but the mission was far from over.

3.5.5 Early Mission Operations

Following liftoff, the spacecraft entered Earth's orbit, and the crew began their early mission operations. They checked the spacecraft's systems, performed necessary course corrections, and prepared for the trans-lunar injection burn that would propel them towards the moon.

Mission Control continued to play a crucial role during this phase of the mission. They monitored the spacecraft's trajectory, calculated the necessary maneuvers, and communicated with the crew to ensure that everything was going according to plan. The countdown had transitioned from the launch phase to the early mission operations phase, and the focus now shifted to the journey to the moon.

As the spacecraft hurtled through space, the crew settled into their routine, conducting experiments, monitoring their health, and preparing for the momentous event that awaited them on the lunar surface. The countdown had brought them to this point, and the world watched with bated breath as the Apollo 11 mission continued its historic journey.

3.6 Launch and Early Mission Operations

The momentous day had finally arrived. After years of preparation and training, the Apollo 11 mission was ready for liftoff. On July 16, 1969, at 9:32 a.m. Eastern Daylight Time, the Saturn V rocket carrying the Apollo 11 spacecraft blasted off from Launch Pad 39A at the Kennedy Space Center in Florida. The world held its breath as the spacecraft soared into the sky, carrying three brave astronauts on a journey that would change the course of history.

3.6.1 Liftoff and Ascent

As the engines roared to life, the immense power of the Saturn V rocket propelled the spacecraft into the air. The ground shook beneath the spectators' feet as they watched the fiery trail of the rocket ascend towards the heavens. Inside the spacecraft, Neil Armstrong, Buzz Aldrin, and Michael Collins felt the intense vibrations and heard the thunderous roar of the engines.

The first few minutes of the ascent were critical. The astronauts had to endure the immense forces of gravity as the rocket accelerated towards space. The spacecraft's systems were closely monitored by the flight controllers in Mission Control, ensuring that everything was functioning as planned. Any deviation from the expected trajectory could have disastrous consequences.

3.6.2 Translunar Injection and Course Correction

Once the spacecraft reached a certain altitude and speed, it was time for the next crucial maneuver: Translunar Injection (TLI). This maneuver involved firing the spacecraft's engines to increase its velocity and send it on a trajectory towards the Moon. The astronauts and the flight controllers in Mission Control worked together to execute this critical maneuver with precision.

After TLI, the spacecraft was on its way to the Moon. However, small course corrections were still necessary to ensure a precise lunar landing. These course corrections were performed by firing the spacecraft's engines at specific times and angles, carefully calculated by the flight controllers. These maneuvers required precise calculations and coordination between the astronauts and Mission Control.

3.6.3 Early Mission Operations

As the spacecraft journeyed towards the Moon, the astronauts settled into their routine. They conducted various experiments, monitored the spacecraft's systems, and communicated with Mission Control. The flight controllers in Mission Control closely monitored the spacecraft's trajectory, systems, and the astronauts' health.

One of the critical tasks during the early mission operations was the navigation of the spacecraft. The astronauts used a combination of celestial navigation and onboard computer systems to determine their position and make any necessary adjustments to their trajectory. This was crucial to ensure that the spacecraft was on the correct path towards the Moon.

Another important aspect of the early mission operations was the monitoring of the spacecraft's systems. The astronauts and the flight controllers worked together to ensure that all systems were functioning properly. They monitored the spacecraft's power, life support systems, communication systems, and other vital components. Any anomalies or malfunctions had to be addressed promptly to ensure the safety of the crew.

Communication between the astronauts and Mission Control was vital throughout the mission. The flight controllers provided the astronauts with updates, instructions, and support. They also relayed messages from the astronauts to their families and the world. The communication systems onboard the spacecraft allowed the astronauts to stay connected with Mission Control despite the vast distance between them.

3.6.4 Challenges and Problem Solving

During the early mission operations, the crew and Mission Control faced several challenges and had to employ their problem-solving skills. One of the most significant challenges was the occurrence of a malfunction in the spacecraft's onboard computer. This computer was essential for navigation and guidance during the lunar landing. The astronauts and the flight controllers worked together to resolve the issue and ensure that the mission could continue.

Another challenge arose when the crew discovered that the carbon dioxide levels in the lunar module were rising to dangerous levels. The flight controllers in Mission Control quickly devised a solution, instructing the astronauts to build a makeshift filter using materials available onboard. This improvised filter successfully removed the excess carbon dioxide and ensured the safety of the crew.

Throughout the early mission operations, the teamwork and collaboration between the astronauts and Mission Control were crucial. The flight controllers provided guidance, support, and expertise, while the astronauts executed the mission objectives and reported back to Mission Control. This seamless coordination and problem-solving mindset were instrumental in overcoming the challenges faced during the mission.

As the spacecraft continued its journey towards the Moon, the world watched in awe and anticipation. The successful launch and early mission operations of Apollo 11 set the stage for the historic lunar landing that was about to take place. The astronauts and the flight controllers had overcome numerous obstacles, demonstrating their unwavering dedication and commitment to the mission's success. The next chapter of the Apollo 11 story would unfold as the spacecraft approached its destination: the Moon.

4 The Journey to the Moon

4.1 Leaving Earth's Atmosphere

Leaving Earth's atmosphere was a crucial and challenging step in the Apollo 11 mission. The astronauts, Neil Armstrong, Buzz Aldrin, and Michael Collins, embarked on a journey that would take them beyond the confines of our planet and into the vastness of space. This section explores the preparations made and the incredible technology that enabled them to leave Earth behind.

4.1.1 The Launch Vehicle: Saturn V

To leave Earth's atmosphere, the Apollo 11 mission relied on the powerful Saturn V launch vehicle. Standing at a staggering height of 363 feet, the Saturn V was the largest and most powerful rocket ever built. It consisted of three stages, each with its own engines, fuel tanks, and specific functions.

The first stage, known as the S-IC, was responsible for the initial boost off the launch pad. Powered by five F-1 engines, it generated an astonishing 7.5 million pounds of thrust. The S-IC stage burned a combination of kerosene and liquid oxygen as propellant.

Once the first stage had completed its task, it was jettisoned, and the second stage, known as the S-II, took over. The S-II stage was powered by five J-2 engines, which burned liquid hydrogen and liquid oxygen. These engines provided the necessary thrust to propel the spacecraft into Earth's orbit.

Finally, the third stage, known as the S-IVB, ignited to push the spacecraft out of Earth's orbit and onto a trajectory towards the Moon. The S-IVB stage had a single J-2 engine and used the same propellant combination as the S-II stage.

4.1.2 The Launch and Ascent

On July 16, 1969, at 9:32 a.m. EDT, the Apollo 11 mission began its historic journey. The Saturn V rocket roared to life, and the immense power of its engines lifted the spacecraft off the launch pad at Kennedy Space Center in Florida. As the rocket ascended, it left behind a trail of smoke and fire, captivating the millions of spectators who watched in awe.

The astronauts inside the Command Module, named Columbia, experienced the intense acceleration as they were propelled into space. Neil Armstrong, Buzz Aldrin, and Michael Collins were well aware of the risks involved in leaving Earth's atmosphere, but their training and preparation had equipped them to handle the challenges that lay ahead.

As the Saturn V rocket climbed higher, it gradually shed its stages, each one serving its purpose and then falling away. The first stage dropped into the Atlantic Ocean, while the second and third stages continued their journey into space, eventually burning up in the Earth's atmosphere or being sent into orbit around the Sun.

4.1.3 Overcoming the Challenges

Leaving Earth's atmosphere presented several challenges that the Apollo 11 mission had to overcome. One of the primary concerns was the immense amount of energy required to break free from Earth's gravitational pull. The Saturn V's powerful engines provided the necessary thrust to achieve this, but precise calculations and engineering were crucial to ensure a successful launch.

Another challenge was the intense heat generated during the ascent. As the rocket climbed, it experienced extreme temperatures due to the friction between the spacecraft and the Earth's atmosphere. To protect the astronauts and the spacecraft from this heat, the Command Module was equipped with a heat shield made of a special material called ablative material. This material absorbed and dissipated the heat, preventing it from reaching the astronauts inside.

Additionally, leaving Earth's atmosphere meant leaving behind the protection of Earth's magnetic field. This exposed the astronauts to the dangers of cosmic radiation, which could potentially harm their health. To mitigate this risk, the spacecraft was equipped with shielding materials and the astronauts were monitored for radiation exposure throughout the mission.

4.1.4 The Journey Begins

As the Saturn V rocket propelled the Apollo 11 spacecraft into space, the astronauts embarked on a journey that would take them to the Moon and back. Leaving Earth's atmosphere was just the first step in this incredible adventure. The next phase would involve docking with the Lunar Module and beginning the long voyage towards the Moon.

The successful departure from Earth's atmosphere was a testament to the ingenuity and dedication of the engineers, technicians, and scientists who worked tirelessly to make the Apollo 11 mission a reality. It was also a testament to the bravery and skill of the astronauts who were willing to venture into the unknown for the sake of exploration and discovery.

As the spacecraft continued its journey, the world held its breath, eagerly awaiting the next milestones in the Apollo 11 mission. The docking with the Lunar Module and the approach to the Moon would bring the astronauts closer to their ultimate goal of landing on the lunar surface and taking that historic first step.

4.2 Docking with the Lunar Module

Once the Apollo 11 spacecraft had successfully left Earth's atmosphere, the next crucial step in the mission was to dock with the Lunar Module (LM). This delicate maneuver was essential for the astronauts to transfer from the Command Module (CM) to the LM, which would ultimately take them to the surface of the moon.

Docking with the LM required precise calculations and careful coordination between the astronauts and mission control. The process involved two main components: the Command Module Pilot (CMP), Michael Collins, and the Lunar Module Pilot (LMP), Buzz Aldrin.

As the CMP, Collins played a vital role in the docking process. He remained in the Command Module, named Columbia, while Armstrong and Aldrin descended to the moon's surface in the LM, named Eagle. Collins had the responsibility of maintaining the orbit of Columbia and ensuring its safe return to Earth.

To initiate the docking procedure, the LM had to separate from the upper stage of the Saturn V rocket. Once separated, the LM had to perform a series of maneuvers to align itself with the Command Module. This alignment was crucial to ensure a successful docking.

The LM's guidance and navigation system played a crucial role in the docking process. It utilized radar and optical sensors to determine the distance and relative velocity between the two spacecraft. This information was relayed to the astronauts and mission control, allowing them to make the necessary adjustments for a precise docking.

As the LM approached the Command Module, Armstrong and Aldrin had to carefully maneuver the spacecraft to align its docking probe with the corresponding drogue on the Command Module. The docking probe was a retractable arm located on the front of the LM, while the drogue was a cone-shaped receptacle on the Command Module.

The docking probe and drogue were designed to fit together with precision, creating a secure connection between the two spacecraft. Once the probe was inserted into the drogue, latches on both sides engaged, ensuring a firm attachment.

The docking process required a high level of coordination and communication between the astronauts and mission control. Armstrong and Aldrin relied on instructions from mission control to guide them through the docking procedure. Mission control closely monitored the progress and provided real-time feedback to the astronauts.

The successful docking of the LM with the Command Module marked a significant milestone in the Apollo 11 mission. It demonstrated the precision and accuracy of the spacecraft's guidance and navigation systems, as well as the skill and expertise of the astronauts.

With the LM securely docked to the Command Module, the astronauts could now transfer from one spacecraft to the other. This transfer was necessary for Armstrong and Aldrin to descend to the lunar surface in the LM. Collins remained in the Command Module, orbiting the moon until their return.

The transfer process involved opening the hatches between the two spacecraft and moving through a small tunnel connecting them. The astronauts had to carefully navigate through the tunnel while wearing their bulky spacesuits, ensuring they didn't damage any equipment or disturb the delicate balance of the spacecraft.

Once inside the LM, Armstrong and Aldrin prepared for their descent to the lunar surface. They checked the systems, reviewed the landing procedures, and made final preparations for their historic moonwalk. The successful docking and transfer set the stage for the next phase of the Apollo 11 mission – landing on the moon.

The docking with the Lunar Module was a critical step in the Apollo 11 mission. It required precise calculations, careful coordination, and the expertise of the astronauts and mission control. The successful docking demonstrated the capabilities of the spacecraft and paved the way for Armstrong and Aldrin to take their historic first steps on the moon.

4.3 The Long Voyage

As the Apollo 11 spacecraft left Earth's orbit, the astronauts embarked on a long and arduous journey towards the moon. This leg of the mission was crucial, as it required precise calculations, careful navigation, and constant monitoring of the spacecraft's systems. The long voyage to the moon would test the endurance and skill of the astronauts, as well as the capabilities of the spacecraft itself.

4.3.1 Navigating the Void

Once the Apollo 11 spacecraft had successfully docked with the lunar module, the crew bid farewell to the command module and began their journey towards the moon. As they ventured further away from Earth, the vastness of space became apparent. The astronauts were surrounded by darkness, with only the distant stars and the pale glow of Earth as their companions.

Navigating through space was a complex task. The astronauts relied on a combination of celestial navigation and precise calculations to ensure they stayed on course. They used star sightings and the position of the sun to determine their location and trajectory. The onboard guidance systems, such as the Apollo Guidance Computer, played a crucial role in calculating and adjusting the spacecraft's path.

4.3.2 Life in Space

Living and working in the confined space of the Apollo spacecraft presented its own set of challenges. The astronauts had to adapt to a microgravity environment, where everyday tasks required careful planning and coordination. Simple actions like eating, sleeping, and using the restroom required adjustments to account for the lack of gravity.

To maintain their physical health, the astronauts followed a strict exercise regimen. They used specially designed equipment to simulate the effects of gravity and prevent muscle and bone loss. Regular exercise was essential to ensure they would be able to perform their duties on the lunar surface.

4.3.3 Communication and Contact

During the long voyage to the moon, communication with mission control and the rest of the world was vital. The astronauts relied on a network of tracking stations located around the globe to maintain contact with Earth. These stations, operated by NASA and international partners, tracked the spacecraft's position and relayed messages between the astronauts and mission control.

The communication system onboard the spacecraft allowed the astronauts to stay in constant contact with mission control. They could receive updates, instructions, and support from the team on the ground. This real-time communication was crucial for troubleshooting any issues that arose during the mission.

4.3.4 The Psychological Toll

The journey to the moon was not only physically demanding but also psychologically challenging for the astronauts. They were isolated from the rest of humanity, confined to a small spacecraft, and faced with the uncertainty of what lay ahead. The weight of their mission and the responsibility they carried added to the psychological strain.

To help cope with the psychological challenges, the astronauts relied on each other for support. They formed a close bond and worked as a team, relying on their training and camaraderie to overcome any difficulties they encountered. Additionally, they had access to a variety of resources, including books, music, and personal items, to provide a sense of comfort and familiarity.

4.3.5 The Fragility of Life

As the astronauts journeyed through space, they were acutely aware of the fragility of life. They were surrounded by the vastness of the universe, with only the thin walls of their spacecraft protecting them from the harsh vacuum of space. Any malfunction or failure could have catastrophic consequences.

The crew and mission control were constantly vigilant, monitoring the spacecraft's systems and addressing any issues that arose. They had to be prepared for any eventuality, from minor malfunctions to life-threatening emergencies. The astronauts underwent extensive training to handle various scenarios and were equipped with the knowledge and tools to respond to emergencies.

4.3.6 A Test of Endurance

The long voyage to the moon tested the endurance of the astronauts both physically and mentally. They had to maintain focus and concentration for extended periods, ensuring the spacecraft stayed on course and all systems operated smoothly. The monotony of the journey was broken only by occasional course corrections and routine tasks.

Despite the challenges, the crew of Apollo 11 remained committed to their mission. They understood the historic significance of their journey and the impact it would have on future generations. Their determination and resilience propelled them forward, inching closer to their ultimate destination.

As the Apollo 11 spacecraft continued its journey towards the moon, the anticipation and excitement grew. The astronauts were on the cusp of achieving what had once seemed impossible - landing humans on the lunar surface. The long voyage was nearing its end, and the next chapter of the Apollo 11 mission was about to unfold.

4.4 Approaching the Moon

As the Apollo 11 spacecraft continued its journey towards the moon, the anticipation and excitement among the crew and mission control grew exponentially. The astronauts, Neil Armstrong, Buzz Aldrin, and Michael Collins, were now just days away from achieving what seemed impossible not too long ago - landing on the moon.

4.4.1 Calculating the Trajectory

Approaching the moon required precise calculations and careful navigation. The mission control team, led by flight director Gene Kranz, worked tirelessly to ensure that the spacecraft was on the correct trajectory. They relied on the expertise of the black women mathematicians, Katherine Johnson, Dorothy Vaughan, and Mary Jackson, who had played a crucial role in the Apollo program.

Using their mathematical skills, these remarkable women calculated the trajectory that would guide the spacecraft towards the moon. Their calculations took into account various factors such as the moon's gravitational pull, the spacecraft's velocity, and the precise timing of the maneuvers required for a successful landing.

4.4.2 Lunar Orbit Insertion

As the spacecraft approached the moon, a critical maneuver known as Lunar Orbit Insertion (LOI) had to be executed. LOI involved firing the spacecraft's engines to slow it down and allow the moon's gravity to capture it into lunar orbit.

On July 19, 1969, as the world held its breath, the Apollo 11 crew initiated the LOI burn. The engines roared to life, and the spacecraft decelerated, gradually entering into orbit around the moon. This crucial maneuver was flawlessly executed, thanks to the meticulous planning and precise calculations carried out by the mission control team and the astronauts.

4.4.3 Lunar Module Separation

With the spacecraft now in lunar orbit, the next step was to separate the Lunar Module (LM), named "Eagle," from the Command Module (CM), named "Columbia." The LM would carry Armstrong and Aldrin to the lunar surface while Collins remained in orbit aboard the CM.

On July 20, 1969, Armstrong and Aldrin entered the LM and prepared for separation. As they bid farewell to Collins, they knew that the success of their mission depended on their ability to safely land the LM on the moon's surface and return to the CM for the journey back to Earth.

4.4.4 Descent and Landing

As the LM descended towards the moon's surface, Armstrong and Aldrin relied on their training and the guidance provided by mission control. The tension in the LM was palpable as they navigated through a field of boulders and craters, searching for a suitable landing site.

With just seconds of fuel remaining, Armstrong skillfully piloted the LM to a safe landing in the Sea of Tranquility. The world erupted in celebration as Armstrong's famous words echoed through the spacecraft and across the globe, "Houston, Tranquility Base here. The Eagle has landed."

4.4.5 The Final Approach

While the world rejoiced at the successful landing, the mission was far from over. Armstrong and Aldrin had to rest and prepare for their historic moonwalk. They meticulously checked their equipment, ensuring that everything was in working order for their descent onto the lunar surface.

Meanwhile, Collins orbited the moon, providing crucial communication and support to his fellow astronauts. He marveled at the sight of the moon's desolate landscape, knowing that his companions were about to embark on a journey that would forever change the course of human history.

4.4.6 The Lunar Surface Beckons

As the hours ticked by, Armstrong and Aldrin made their final preparations. They donned their spacesuits, checked their life support systems, and depressurized the LM. The moment had come for them to take humanity's first steps on another celestial body.

With the world watching and holding its breath, Armstrong descended the ladder of the LM and set foot on the moon's surface. His words, "That's one small step for man, one giant leap for mankind," resonated with people around the world, capturing the magnitude of the achievement.

Aldrin soon joined Armstrong on the lunar surface, and together they conducted experiments, collected samples, and planted the American flag. They marveled at the desolate beauty of the moon and the vastness of space that stretched out before them.

4.4.7 Awe and Wonder

As Armstrong and Aldrin explored the lunar surface, their experiences were shared with the world through live television broadcasts. People from all walks of life watched in awe and wonder as they witnessed the triumph of human ingenuity and determination.

The images transmitted from the moon's surface showed the astronauts bouncing across the lunar landscape, leaving footprints in the fine lunar dust. They conducted experiments to better understand the moon's geology and its potential for future exploration.

4.4.8 Returning to the Command Module

After spending approximately two and a half hours on the lunar surface, Armstrong and Aldrin returned to the LM. They prepared for their ascent back to the CM, leaving behind scientific instruments and equipment that would continue to transmit valuable data to Earth.

With the LM safely docked with the CM, the astronauts transferred back into the Command Module, reuniting with Collins. The three astronauts prepared for their journey back to Earth, knowing that their mission had been an unprecedented success.

4.4.9 The Journey Continues

As the Apollo 11 spacecraft left the moon's orbit, the world celebrated the remarkable achievement of landing humans on another celestial body. The journey back to Earth was filled with reflection and anticipation, as the astronauts contemplated the impact of their mission and the legacy it would leave behind.

Little did they know that their voyage to the moon would inspire generations to come, igniting a passion for space exploration and pushing the boundaries of human knowledge. The Apollo 11 mission was a testament to the power of human determination, teamwork, and the indomitable spirit of exploration.

As the spacecraft hurtled through space, the astronauts and mission control knew that their mission was far from over. They had successfully completed the first leg of their journey, but there were still challenges to overcome and a world waiting to welcome them home.

5 Landing on the Moon

5.1 Preparing for Descent

As the Apollo 11 spacecraft approached the Moon, the crew inside the Command Module, consisting of Neil Armstrong, Buzz Aldrin, and Michael Collins, prepared themselves for the crucial phase of the mission: the descent to the lunar surface. This phase required meticulous planning, precise calculations, and a deep understanding of the lunar environment.

5.1.1 Calculating the Descent Trajectory

One of the key challenges in preparing for the descent was calculating the trajectory that would safely guide the Lunar Module, named Eagle, to the surface of the Moon. This task fell to the team of flight controllers and mathematicians in Mission Control, led by Flight Director Gene Kranz. They worked tirelessly to ensure that the descent trajectory would bring the Lunar Module to the intended landing site in the Sea of Tranquility.

To calculate the trajectory, the team took into account various factors, such as the Moon's gravitational pull, the spacecraft's velocity, and the position of the landing site. They used complex mathematical equations and computer simulations to determine the precise path that the Lunar Module would follow during descent.

5.1.2 Simulating the Descent

To further prepare for the descent, the astronauts underwent extensive training in simulators that replicated the conditions they would encounter during the actual landing. These simulators allowed them to practice controlling the Lunar Module's descent engine and maneuvering it towards the lunar surface.

The simulations were designed to mimic the low-gravity environment of the Moon, as well as the challenges posed by the lack of atmosphere. The astronauts had to learn how to make precise adjustments to the Lunar

Module's descent engine throttle and attitude to ensure a smooth and controlled landing.

5.1.3 Monitoring Systems and Instruments

During the descent, the astronauts relied on a multitude of systems and instruments to guide them safely to the lunar surface. These systems included radar altimeters, which measured the distance between the Lunar Module and the Moon's surface, and the guidance computer, which provided real-time data on the spacecraft's position and velocity.

The flight controllers in Mission Control closely monitored these systems and instruments, ensuring that they were functioning properly and providing accurate information to the astronauts. Any anomalies or deviations from the expected values were immediately addressed, and the flight controllers worked together with the astronauts to troubleshoot and resolve any issues.

5.1.4 Final Preparations

As the Lunar Module descended towards the Moon, the astronauts made final preparations for the landing. Neil Armstrong, as the mission commander, took the lead in overseeing these preparations. He ensured that all systems were functioning correctly, and that the Lunar Module was on the correct trajectory for landing.

Buzz Aldrin, the Lunar Module pilot, closely monitored the descent and communicated with Mission Control, providing updates on the spacecraft's status and relaying any important information to the flight controllers. Michael Collins, in the Command Module orbiting above, maintained communication with both the Lunar Module and Mission Control, acting as a vital link between the two.

The tension inside the spacecraft was palpable as the astronauts prepared to make history. They knew that the success of the mission hinged on their ability to land the Lunar Module safely on the Moon's surface. The culmination of years of planning, training, and hard work was about to be put to the ultimate test.

With each passing second, the Lunar Module descended closer to the Moon. The astronauts focused intently on their instruments, their training guiding their every move. The world held its breath as the Eagle prepared to make its historic landing.

Word Count: 582

5.2 The Eagle Has Landed

As the Apollo 11 spacecraft descended towards the lunar surface, the tension in Mission Control was palpable. The culmination of years of planning, training, and technological advancements had led to this moment. The fate of the mission and the success of the United States in the Space Race rested on the shoulders of the three brave astronauts aboard the Lunar Module, Neil Armstrong, Buzz Aldrin, and Michael Collins.

5.2.1 The Final Approach

On July 20, 1969, the Lunar Module, named "Eagle," separated from the Command Module, "Columbia," piloted by Michael Collins. As the Eagle began its descent towards the Moon's surface, the astronauts inside were acutely aware of the risks and challenges they faced. The lunar module's guidance computer, with its limited processing power, had to navigate the treacherous terrain and find a suitable landing site.

The tension in Mission Control was palpable as the flight controllers monitored every aspect of the descent. The astronauts' heart rates increased as they descended further, their eyes fixed on the lunar landscape rapidly approaching beneath them. The lunar module's thrusters fired, adjusting its trajectory and slowing its descent. The mission's success hinged on the precise execution of this critical maneuver.

5.2.2 A Nerve-Wracking Descent

As the Eagle descended, Neil Armstrong, the mission commander, took manual control of the lunar module. The automated landing system had brought them close to a large crater, and Armstrong had to make a split-second decision to override the computer's guidance and find a safer landing spot. With only seconds of fuel remaining, Armstrong skillfully maneuvered the lunar module to a more suitable location.

The tension in Mission Control reached its peak as the famous words were transmitted from the lunar module to Earth, "The Eagle has landed." The room erupted in cheers and applause as the realization sank in that humanity had achieved the impossible. The first humans had successfully landed on the Moon.

5.2.3 A Historic Moment

Neil Armstrong and Buzz Aldrin prepared for their historic moonwalk. They donned their bulky spacesuits and depressurized the lunar module. As the hatch opened, Armstrong descended the ladder and set foot on the lunar surface, uttering the now-famous words, "That's one small step for man, one giant leap for mankind."

Buzz Aldrin soon joined Armstrong on the surface, and together they planted the American flag, symbolizing the United States' achievement in the Space Race. They conducted experiments, collected samples, and took photographs, all while being watched by millions of people around the world. The images and videos transmitted back to Earth captured the imagination of people everywhere, inspiring generations to dream of exploring the cosmos.

5.2.4 The Unsung Heroes

While Armstrong and Aldrin were making history on the lunar surface, Michael Collins orbited the Moon in the Command Module, Columbia. Collins played a crucial role in the mission, ensuring the safe return of his fellow astronauts. Alone in the Command Module, he patiently waited for their return, providing vital communication and support.

Back on Earth, the dedicated team in Mission Control worked tirelessly to ensure the success of the mission. Flight directors, such as Gene Kranz, led the team of flight controllers, monitoring every aspect of the mission and making critical decisions in real-time. The expertise and dedication of these individuals were instrumental in the safe landing and return of the Apollo 11 crew.

5.2.5 A Moment of Reflection

As Armstrong and Aldrin explored the lunar surface, they marveled at the desolate beauty of the Moon. They collected rock samples, conducted experiments, and documented their experiences. But amidst the awe and wonder, they also took a moment to reflect on the significance of their achievement.

They recognized that their successful landing on the Moon was not just a victory for the United States but for all of humanity. It was a testament to the power of human ingenuity, determination, and collaboration. The Apollo 11 mission represented the pinnacle of scientific and technological achievement, pushing the boundaries of what was thought possible.

The landing on the Moon was a moment of unity and inspiration for people around the world. It showcased the potential of human exploration and the boundless possibilities that lie beyond our home planet. The images and stories of the Apollo 11 mission captivated the world, leaving an indelible mark on history and inspiring future generations to reach for the stars.

As Armstrong and Aldrin prepared to leave the lunar surface and return to the Command Module, they knew that their mission was far from over. The journey back to Earth would present its own set of challenges and uncertainties. But for that brief moment on the Moon, they had achieved the impossible and forever changed the course of human history.

5.3 Stepping onto the Lunar Surface

After a long and arduous journey, the moment had finally arrived. On July 20, 1969, Neil Armstrong and Buzz Aldrin prepared to take humanity's first steps on the lunar surface. The world held its breath as the two astronauts made their way to the Lunar Module's hatch.

5.3.1 The Egress Procedure

Stepping onto the lunar surface was not as simple as opening a door and walking out. The astronauts had to follow a carefully planned egress procedure to ensure their safety and the success of the mission. They had to don their Extravehicular Mobility Units (EMUs), which were specially designed spacesuits for the lunar environment. These suits provided them with protection from the harsh conditions of the moon, including extreme temperatures and the lack of atmosphere.

Once suited up, Armstrong and Aldrin depressurized the Lunar Module's cabin and opened the hatch. As they descended the ladder, they had to be cautious of their movements, as the lunar gravity was only one-sixth that of Earth's. Every step had to be deliberate and calculated to maintain balance and stability.

5.3.2 The First Step

As Neil Armstrong reached the bottom of the ladder, he paused for a moment to assess the lunar surface. The world watched in anticipation as he uttered the now-famous words, "That's one small step for man, one giant leap for mankind." With those words, Armstrong became the first human to set foot on the moon.

Armstrong's first step was not just a personal achievement; it was a monumental moment for all of humanity. It represented the culmination of years of hard work, dedication, and scientific progress. The Apollo 11 mission had achieved what was once thought impossible, and the world celebrated this historic milestone.

5.3.3 The Lunar Environment

Stepping onto the lunar surface was a surreal experience for Armstrong and Aldrin. The moon's landscape was unlike anything they had ever seen before. The surface was covered in fine, powdery dust, which clung to their spacesuits and made their movements slightly more challenging.

The astronauts marveled at the desolate beauty of the moon. The sky above them was pitch black, devoid of any atmosphere or clouds. The stars shone brightly, unobstructed by Earth's atmosphere. The silence was profound, as there was no air to carry sound waves. It was a completely alien environment, and yet, Armstrong and Aldrin were there, representing all of humanity.

5.3.4 Collecting Samples and Conducting Experiments

While on the lunar surface, Armstrong and Aldrin had important scientific tasks to complete. They collected samples of the moon's soil and rocks, using specially designed tools and containers. These samples would provide valuable insights into the moon's geology and history, as well as help scientists better understand the origins of our solar system.

In addition to sample collection, the astronauts also conducted various experiments. They deployed scientific instruments to measure the moon's seismic activity, temperature, and other environmental factors. These experiments would contribute to our understanding of the moon and its potential for future exploration.

5.3.5 Planting the American Flag

As a symbol of American achievement, Armstrong and Aldrin planted the American flag on the lunar surface. This act represented not only the success of the Apollo 11 mission but also the United States' victory in the Space Race against the Soviet Union. The flag stood tall and proud, a testament to human ingenuity and the spirit of exploration.

5.3.6 The Lunar Walk and Communication

Throughout their time on the lunar surface, Armstrong and Aldrin maintained constant communication with Mission Control in Houston. They provided updates on their activities, relayed scientific data, and received instructions for their tasks. This communication was vital for the success and safety of the mission.

Armstrong and Aldrin's lunar walk lasted approximately two and a half hours. During this time, they explored the immediate vicinity of the Lunar Module, taking photographs and documenting their experiences. They moved with caution, aware of the limited time they had on the surface and the importance of completing their objectives.

5.3.7 Returning to the Lunar Module

As their time on the lunar surface came to an end, Armstrong and Aldrin made their way back to the Lunar Module. They carefully climbed the ladder, ensuring that no lunar dust contaminated the interior of the spacecraft. Once inside, they removed their spacesuits and prepared for the journey back to Earth.

Stepping onto the lunar surface was a defining moment in human history. It marked the realization of President Kennedy's vision and the culmination of years of scientific and technological advancements. Neil Armstrong and Buzz Aldrin's footsteps on the moon inspired generations to dream big and pushed the boundaries of what humanity could achieve. The Apollo 11 mission was a testament to the power of human determination, ingenuity, and the unwavering spirit of exploration.

5.4 Exploring the Moon

After the historic landing of the Apollo 11 Lunar Module, known as the Eagle, on the surface of the moon, the astronauts had a monumental task ahead of them. Their mission was not only to set foot on the lunar surface but also to explore and gather valuable scientific data. This section will delve into the incredible journey of exploration undertaken by Neil Armstrong and Buzz Aldrin as they ventured into the unknown.

5.4.1 The Lunar Landscape

As Neil Armstrong took his first step onto the moon's surface, he described it as "one small step for man, one giant leap for mankind." The astronauts were greeted by a desolate and barren landscape, unlike anything they had ever seen before. The moon's surface was covered in a fine layer of gray dust, known as regolith, which made walking a bit challenging. The regolith was the result of billions of years of meteorite impacts, creating a powdery texture that clung to the astronauts' spacesuits and equipment.

5.4.2 Collecting Samples

One of the primary objectives of the Apollo 11 mission was to collect samples of the moon's surface and bring them back to Earth for analysis. The astronauts used specially designed tools, such as the rock hammer and scoop, to gather rocks and soil samples. These samples would provide scientists with valuable insights into the moon's composition and history.

Neil Armstrong and Buzz Aldrin carefully selected a variety of rocks from different locations to ensure a diverse collection. They also documented the exact location of each sample, taking photographs and making detailed notes. The samples were stored in sealed containers to prevent any contamination and were later returned to Earth for extensive analysis.

5.4.3 Setting Up Experiments

In addition to collecting samples, the astronauts also deployed several scientific experiments on the lunar surface. These experiments were designed to gather data on various aspects of the moon's environment and its interaction with space.

One of the experiments was the Passive Seismic Experiment, which consisted of seismometers placed on the moon's surface to detect any seismic activity. This experiment aimed to provide insights into the moon's internal structure and the presence of moonquakes.

Another experiment was the Lunar Laser Ranging Retroreflector, which involved placing a series of retroreflectors on the moon's surface. These retroreflectors reflected laser beams sent from Earth, allowing scientists to precisely measure the distance between the Earth and the moon. This experiment provided valuable data for studying the moon's orbit and gravitational interactions.

5.4.4 Exploring the Area

As Neil Armstrong and Buzz Aldrin explored the moon's surface, they moved away from the Lunar Module, conducting experiments and documenting their findings. They carefully traversed the uneven terrain, hopping from one spot to another due to the moon's lower gravity.

The astronauts used a variety of tools and equipment to aid in their exploration. They carried a portable life support system, which provided them with oxygen and regulated their body temperature. They also had a Hasselblad camera to capture high-resolution photographs of the lunar landscape and their activities.

During their exploration, the astronauts encountered several challenges. The moon's surface was covered in fine dust, which had a tendency to cling to their spacesuits and equipment. This dust posed a risk of clogging the equipment and hindering their movements. However, the astronauts managed to adapt and overcome these challenges, showcasing their exceptional training and resourcefulness.

5.4.5 Discoveries and Findings

The exploration of the moon's surface yielded numerous discoveries and findings that significantly contributed to our understanding of the moon and its formation. The analysis of the collected samples revealed that the moon's surface was primarily composed of basalt, a type of volcanic rock. This finding provided evidence of past volcanic activity on the moon.

Scientists also discovered that the moon's surface had been bombarded by meteorites for billions of years, resulting in the creation of numerous craters. The study of these craters helped scientists understand the history of impacts in the solar system and their effects on planetary bodies.

Additionally, the experiments deployed by the astronauts provided valuable data on the moon's seismic activity, gravitational field, and its interaction with the Earth. These findings helped scientists refine their models of the moon's formation and evolution.

Conclusion

The exploration of the moon by Neil Armstrong and Buzz Aldrin during the Apollo 11 mission was a remarkable achievement for humanity. Their footsteps on the lunar surface marked a pivotal moment in history and opened the door to further exploration of our solar system.

The samples collected and experiments conducted during the mission continue to provide scientists with valuable insights into the moon's composition, history, and its relationship with Earth. The discoveries made during the Apollo 11 mission laid the foundation for future lunar missions and paved the way for our continued exploration of space.

As the astronauts prepared to leave the moon's surface and return to Earth, they carried with them not only the physical samples but also the knowledge and inspiration gained from their extraordinary journey. The legacy of Apollo 11 and the exploration of the moon will forever be etched in the annals of human achievement, serving as a testament to the indomitable spirit of exploration and the boundless possibilities of the human mind.

6 The Return Journey

6.1 Saying Goodbye to the Moon

As the Apollo 11 mission neared its end, the astronauts aboard the spacecraft prepared to bid farewell to the moon, their temporary home for the past few days. The time had come for them to leave the lunar surface and begin their journey back to Earth. Saying goodbye to the moon was a bittersweet moment for the crew of Apollo 11, as they had achieved what seemed impossible just a few days earlier.

On July 20, 1969, Neil Armstrong and Buzz Aldrin had become the first humans to set foot on the moon. Their historic moonwalk had captivated the world and marked a significant milestone in human history. But now, it was time to wrap up their mission and return home.

Before leaving the moon, the astronauts had several tasks to complete. They carefully documented their activities, collecting samples of lunar soil and rocks, and taking photographs of the lunar surface. These samples and images would provide valuable scientific data and help researchers gain a deeper understanding of the moon's composition and history.

As they prepared to depart, the astronauts took a moment to reflect on their experiences. Neil Armstrong, the first man to walk on the moon, famously described the lunar surface as "one small step for man, one giant leap for mankind." These words captured the significance of their achievement and the collective effort that had brought them to this point.

Buzz Aldrin, the lunar module pilot, also took a moment to commemorate the occasion. He planted the American flag on the moon's surface, symbolizing the United States' successful mission and its commitment to space exploration. The flag would remain on the moon as a testament to the human spirit of exploration and discovery.

After completing their tasks, the astronauts returned to the lunar module, known as the Eagle, to prepare for their departure. They carefully sealed the samples they had collected and secured them for the journey back to Earth. The lunar module had served them well during their time on the moon, but now it was time to leave it behind.

As the lunar module prepared for liftoff, Michael Collins, the command module pilot, orbited above in the command module, Columbia. He anxiously awaited the return of his fellow astronauts, knowing that their safe departure from the moon was crucial for the success of the mission.

With everything in order, the lunar module fired its engines, propelling Armstrong and Aldrin back into space. As they left the moon's surface, they looked back at the desolate landscape they were leaving behind. The sight of the moon, with its craters and vast expanse, was a reminder of the challenges they had overcome and the triumphs they had achieved.

Once the lunar module docked with the command module, the three astronauts reunited, and the lunar module was jettisoned. They were now ready to begin their journey back to Earth. The return journey would be long and arduous, but the crew was prepared for the challenges that lay ahead.

As they left the moon behind, the astronauts couldn't help but feel a sense of awe and wonder at what they had accomplished. They had not only fulfilled President Kennedy's vision of landing a man on the moon but had also inspired generations to come. Their bravery and determination had pushed the boundaries of human exploration and opened up new possibilities for the future.

The return journey was not without its own set of challenges. The crew had to navigate through space, perform critical maneuvers, and ensure the safe reentry of the command module into Earth's atmosphere. Mission control played a vital role in guiding them through these final stages of the mission, providing support and guidance every step of the way.

Finally, on July 24, 1969, the command module splashed down in the Pacific Ocean, bringing an end to the historic Apollo 11 mission. The crew was safely recovered by the USS Hornet, and they were immediately placed in quarantine to prevent the potential spread of any lunar pathogens.

The successful completion of the Apollo 11 mission marked a turning point in human history. It demonstrated the power of human ingenuity, determination, and collaboration. The legacy of Apollo 11 would extend far beyond the moon landing itself, inspiring future generations to reach for the stars and explore the unknown.

As the crew of Apollo 11 said goodbye to the moon, they left behind a lasting legacy. Their achievements would forever be etched in the annals of history, reminding us of what can be accomplished when we dare to dream and work together to achieve the impossible. The journey to the moon was over, but the impact of Apollo 11 would continue to shape the future of space exploration and inspire generations to come.

6.2 Reentry and Splashdown

After spending several days on the lunar surface, Neil Armstrong, Buzz Aldrin, and Michael Collins prepared for their return journey to Earth. The success of their mission relied not only on their ability to land on the moon but also on their safe return. Reentry and splashdown were critical phases of the Apollo 11 mission, and meticulous planning and precise execution were essential to ensure the astronauts' safe return.

6.2.1 Preparing for Reentry

As the lunar module, named Eagle, lifted off from the moon's surface and docked with the command module, named Columbia, the astronauts transferred back to the command module. Once inside, they prepared for the journey back to Earth. The command module was equipped with a heat shield, which would protect the astronauts from the intense heat generated during reentry into Earth's atmosphere.

Before reentry, the astronauts had to jettison the lunar module, as it was no longer needed. The separation of the lunar module was a critical step, and any malfunction could have jeopardized the entire mission. However, the separation was executed flawlessly, and the lunar module was left behind as the command module prepared for its descent.

6.2.2 Reentry into Earth's Atmosphere

On July 24, 1969, as Apollo 11 approached Earth, the command module began its descent into the atmosphere. Reentry was a highly intense and dangerous phase of the mission. The command module experienced extreme temperatures, reaching up to 5,000 degrees Fahrenheit, as it plowed through the Earth's atmosphere at a speed of approximately 25,000 miles per hour.

The heat shield played a crucial role in protecting the astronauts from the searing heat. Made of a material called ablator, the heat shield absorbed and dissipated the heat, preventing it from reaching the command module's interior. The astronauts could feel the intense heat outside, but inside the module, they remained safe and secure.

As the command module descended, it experienced tremendous deceleration forces. The astronauts felt the pressure pushing them back into their seats, similar to the force they experienced during liftoff. However, this time, the force was in the opposite direction, slowing down the module and allowing it to safely reenter the Earth's atmosphere.

6.2.3 Parachute Deployment and Splashdown

As the command module descended further into the atmosphere, the astronauts deployed a series of parachutes to slow down its descent. The parachutes were crucial in reducing the module's speed and ensuring a safe landing in the ocean.

The first parachute, known as the drogue chute, was deployed at an altitude of approximately 21,000 feet. Its purpose was to stabilize the module and slow it down further. Once the drogue chute had done its job, the main parachutes were deployed. These massive parachutes, measuring over 80 feet in diameter, slowed the module's descent to a safe speed.

As the command module descended under the main parachutes, it approached the splashdown zone in the Pacific Ocean. The recovery team, consisting of helicopters, aircraft, and naval vessels, was stationed nearby, ready to retrieve the astronauts and the module.

Finally, on July 24, 1969, at 12:50 p.m. Eastern Daylight Time, the command module splashed down in the Pacific Ocean, approximately 900 miles southwest of Hawaii. The splashdown was a remarkable achievement, marking the successful completion of the Apollo 11 mission.

6.2.4 Recovery and Quarantine

As soon as the command module splashed down, the recovery team sprang into action. Helicopters hovered above the module, and divers approached to assist the astronauts in exiting the module and boarding the recovery vessel, the USS Hornet.

Once aboard the USS Hornet, the astronauts underwent a series of medical checks to ensure their well-being. They were also required to spend several weeks in quarantine to prevent the potential spread of any lunar pathogens. This precautionary measure was taken because scientists were unsure if the moon harbored any harmful microorganisms that could pose a threat to Earth's ecosystem.

During their quarantine period, the astronauts were closely monitored and underwent various tests to ensure they remained healthy. The quarantine was lifted on August 10, 1969, after it was determined that the astronauts posed no risk to public health.

6.2.5 Celebrating the Historic Achievement

The successful reentry and splashdown of Apollo 11 marked the end of an extraordinary journey. The entire world celebrated the safe return of the astronauts, recognizing their remarkable achievement in landing on the moon and returning home.

Upon their arrival in Hawaii, the astronauts were greeted with a hero's welcome. They were honored with parades, ceremonies, and accolades from people around the globe. The Apollo 11 mission had captivated the world's attention, and the safe return of the astronauts brought a sense of relief and pride to everyone involved.

The reentry and splashdown of Apollo 11 not only marked the end of a historic mission but also paved the way for future space exploration. The knowledge gained from this mission would shape the future of space travel and inspire generations to come. The safe return of the astronauts was a testament to the dedication, skill, and teamwork of everyone involved in the Apollo program.

6.3 Recovery and Quarantine

After the successful completion of their historic mission on the moon, the Apollo 11 astronauts faced a new set of challenges upon their return to Earth. Recovery and quarantine were crucial steps in ensuring the safety of the astronauts and preventing any potential contamination from the lunar surface.

6.3.1 The Importance of Recovery

As the Apollo 11 spacecraft reentered Earth's atmosphere, the world held its breath, anxiously awaiting the safe return of Neil Armstrong, Buzz Aldrin, and Michael Collins. The recovery phase of the mission involved the retrieval of the astronauts and their spacecraft from the Pacific Ocean, where they splashed down after their journey through space.

To facilitate the recovery process, NASA had established a well-coordinated system involving various teams and resources. The primary recovery vessel was the USS Hornet, an aircraft carrier specially equipped to handle the retrieval of the Apollo spacecraft. The Hornet was stationed in the designated recovery area in the Pacific Ocean, ready to spring into action as soon as the astronauts touched down.

6.3.2 Retrieval and Medical Evaluation

Once the Apollo 11 spacecraft splashed down, a team of Navy divers quickly secured the capsule and prepared it for retrieval. Helicopters were then deployed to hoist the astronauts and the capsule onto the deck of the USS Hornet. The recovery team worked efficiently to ensure the safety of the astronauts and the preservation of any potential evidence from the mission.

Upon their arrival on the Hornet, the astronauts were immediately taken to the Mobile Quarantine Facility (MQF), a specially designed trailer that would serve as their temporary home for the next few weeks. The MQF was a sealed environment, ensuring that the astronauts would not come into contact with anyone or anything outside until they had undergone thorough medical evaluations.

6.3.3 Quarantine Procedures

The quarantine period was a critical part of the recovery process, as it aimed to prevent the potential spread of any lunar pathogens or contaminants that the astronauts might have brought back from the moon. The quarantine measures were put in place based on the prevailing scientific understanding at the time, which suggested the possibility of unknown lunar microorganisms.

Inside the Mobile Quarantine Facility, the astronauts had access to a comfortable living space, complete with sleeping quarters, a small kitchen, and a bathroom. They were able to communicate with mission control and their families through a specially designed isolation chamber, which allowed for video and audio transmissions.

During the quarantine period, the astronauts underwent a series of medical tests and examinations to ensure their well-being. They were closely monitored for any signs of illness or infection. The medical team also collected samples from the astronauts, including blood, urine, and saliva, to analyze for potential lunar contaminants.

6.3.4 Lunar Sample Quarantine

In addition to the quarantine of the astronauts, strict measures were taken to isolate and study the lunar samples brought back from the mission. The lunar samples were considered extremely valuable scientific specimens, and their handling required meticulous care to prevent any contamination.

A separate facility, known as the Lunar Receiving Laboratory (LRL), was established at the NASA Manned Spacecraft Center (now Johnson Space Center) in Houston, Texas. The LRL was a highly secure and controlled environment, equipped with specialized equipment and laboratories for the analysis and study of the lunar samples.

Scientists and researchers worked tirelessly at the LRL to examine the lunar samples and unlock the secrets of the moon. The samples were carefully cataloged, analyzed, and distributed to various research institutions around the world for further study. The findings from these studies would contribute significantly to our understanding of the moon's geology, history, and potential for future exploration.

6.3.5 Release from Quarantine

After a quarantine period of approximately three weeks, the Apollo 11 astronauts were finally released from their isolation. On August 10, 1969, they emerged from the Mobile Quarantine Facility, greeted by a jubilant crowd and a hero's welcome. The successful completion of their quarantine marked the final step in their journey from the moon back to Earth.

The recovery and quarantine procedures implemented during the Apollo 11 mission were not only crucial for the safety of the astronauts but also for the protection of Earth's biosphere. These measures demonstrated NASA's commitment to responsible space exploration and set a precedent for future missions to ensure the containment of potential extraterrestrial contaminants.

The recovery and quarantine phase of the Apollo 11 mission showcased the dedication and meticulous planning of the teams involved. It was a testament to the collaborative efforts of NASA, the Navy, and the medical community in ensuring the safe return of the astronauts and the preservation of the scientific integrity of the mission. The successful completion of this phase paved the way for the next chapter in the Apollo 11 story: the celebration of their historic achievement and the lasting legacy of the moon landing.

6.4 Celebrating the Historic Achievement

The successful completion of the Apollo 11 mission marked a monumental moment in human history. The entire world watched in awe as Neil Armstrong took his first steps on the lunar surface, forever etching his name in the annals of space exploration. The achievement of landing on the moon was not only a triumph for the United States but for all of humanity. The celebration that followed the historic achievement was a testament to the collective spirit of mankind and the boundless possibilities of human ingenuity.

6.4.1 A Global Celebration

The news of the successful moon landing spread like wildfire across the globe. People from all walks of life, regardless of nationality or background, joined together in celebrating this extraordinary feat. The achievement of Apollo 11 transcended political boundaries and united the world in a shared sense of wonder and pride. From bustling cities to remote villages, the historic achievement was met with jubilation and a renewed sense of hope for the future.

6.4.2 Honoring the Astronauts

Neil Armstrong, Buzz Aldrin, and Michael Collins became instant heroes and symbols of human achievement. Their names were etched into the history books, and their faces adorned newspapers and television screens around the world. Parades, ceremonies, and public events were held to honor their bravery and dedication. The astronauts were hailed as pioneers and explorers, embodying the spirit of adventure and pushing the boundaries of what was thought possible.

6.4.3 Presidential Recognition

President Richard Nixon played a significant role in recognizing the historic achievement of Apollo 11. He personally greeted the astronauts upon their return to Earth and awarded them the Presidential Medal of Freedom, the highest civilian honor in the United States. In his address to the nation, President Nixon praised the astronauts for their courage and the entire Apollo 11 team for their remarkable achievement. He acknowledged the significance of the mission, not only for the United States but for all of humanity.

6.4.4 Parades and Festivities

Cities across the United States organized parades and festivities to celebrate the successful moon landing. Millions of people lined the streets, waving American flags and cheering as the astronauts passed by in open-top cars. The atmosphere was electric, with a palpable sense of pride and excitement. The celebrations were not limited to the United States alone; countries around the world held their own events to commemorate this historic achievement.

6.4.5 Scientific Conferences and Exhibitions

In addition to the public celebrations, scientific conferences and exhibitions were organized to showcase the scientific and technological advancements made during the Apollo 11 mission. Experts from various fields gathered to discuss the mission's findings and the implications for future space exploration. These conferences served as a platform for scientists and engineers to exchange ideas and collaborate on future missions.

6.4.6 Cultural Impact

The moon landing had a profound impact on popular culture. Books, movies, and songs were inspired by the Apollo 11 mission, capturing the imagination of people around the world. The iconic image of Buzz Aldrin standing on the lunar surface became an enduring symbol of human achievement. The moon landing also sparked a renewed interest in science and space exploration, inspiring a new generation of scientists, engineers, and astronauts.

6.4.7 Legacy of Apollo 11

The legacy of Apollo 11 extends far beyond the historic achievement of landing on the moon. The mission paved the way for future space exploration and scientific discoveries. The technology developed for the Apollo program has found applications in various fields, from medicine to telecommunications. The success of Apollo 11 also served as a catalyst for international cooperation in space exploration, leading to collaborative missions and the establishment of the International Space Station.

6.4.8 Inspiring Future Generations

Perhaps the most significant impact of Apollo 11 was its ability to inspire future generations. The mission showed that with determination, innovation, and teamwork, seemingly insurmountable challenges can be overcome. It ignited a passion for exploration and a curiosity about the universe that continues to drive scientific progress. The astronauts themselves became role models for aspiring scientists and engineers, encouraging young minds to pursue careers in STEM fields.

6.4.9 The Power of Diversity

The celebration of the Apollo 11 mission also highlighted the critical involvement of black women mathematicians, whose contributions were instrumental to the success of the mission. Katherine Johnson, Dorothy Vaughan, and Mary Jackson, known as the "Hidden Figures," played pivotal roles in calculating trajectories, analyzing data, and ensuring the safety of the astronauts. Their remarkable achievements shattered barriers and paved the way for greater diversity and inclusion in the scientific community.

6.4.10 A Lasting Inspiration

The celebration of the Apollo 11 mission continues to this day, serving as a reminder of the indomitable spirit of human exploration. It stands as a testament to what can be achieved when nations come together, pushing the boundaries of knowledge and venturing into the unknown. The historic achievement of Apollo 11 will forever be remembered as a defining moment in human history, inspiring generations to dream big and reach for the stars.

7 The Legacy of Apollo 11

7.1 Scientific Discoveries and Lunar Samples

The Apollo 11 mission not only marked a monumental achievement in human history but also paved the way for significant scientific discoveries and advancements. The scientific objectives of the mission were to gather valuable data and samples from the lunar surface, providing researchers with a wealth of information about the Moon's composition, geology, and history.

7.1.1 Lunar Sample Collection

One of the primary goals of the Apollo 11 mission was to collect lunar samples and bring them back to Earth for analysis. The astronauts, Neil Armstrong and Buzz Aldrin, spent a total of about two and a half hours outside the Lunar Module, conducting experiments and gathering samples. They carefully selected rocks and soil samples from various locations on the lunar surface, ensuring a diverse representation of the Moon's geological history.

The lunar samples collected during the Apollo 11 mission provided scientists with an unprecedented opportunity to study the Moon's composition and formation. These samples offered insights into the Moon's age, its relationship to Earth, and the processes that shaped its surface over billions of years. By analyzing the chemical and isotopic composition of the samples, scientists were able to gain a deeper understanding of the Moon's origins and its role in the early solar system.

7.1.2 Lunar Geology and Volcanism

The study of the lunar samples brought back by Apollo 11 revealed fascinating insights into the Moon's geology and volcanic history. The samples showed evidence of ancient volcanic activity, with basaltic rocks indicating past lava flows on the lunar surface. By analyzing the composition and age of these rocks, scientists were able to reconstruct the timeline of volcanic activity on the Moon and gain a better understanding of its geological evolution.

The presence of impact craters on the lunar surface was another significant discovery made possible by the Apollo 11 mission. The samples collected from these impact sites provided valuable information about the frequency and intensity of meteorite impacts on the Moon. By studying the distribution and characteristics of these impact craters, scientists were able to gain insights into the history of the solar system and the processes that shaped the Moon's surface.

7.1.3 Lunar Water and Ice

One of the most surprising discoveries made by the Apollo 11 mission was the presence of water molecules on the lunar surface. While the Moon was previously thought to be completely dry, the analysis of lunar samples revealed the existence of trace amounts of water. This finding challenged previous assumptions about the Moon's composition and raised intriguing questions about the origin and distribution of water in the solar system.

Further studies conducted on lunar samples from subsequent Apollo missions confirmed the presence of water ice in permanently shadowed regions near the lunar poles. These findings have significant implications for future lunar exploration and the potential for establishing a sustainable human presence on the Moon. The discovery of water on the Moon has opened up possibilities for using it as a resource for future missions, such as providing drinking water, generating oxygen, and producing rocket propellant.

7.1.4 Solar Wind and Cosmic Ray Studies

In addition to collecting lunar samples, the Apollo 11 mission also carried scientific instruments to study the solar wind and cosmic rays. The lunar surface provided an ideal location for these measurements, as it is not shielded by Earth's atmosphere and magnetic field. By analyzing the data collected during the mission, scientists were able to gain insights into the nature of the solar wind and its interaction with the lunar surface.

The solar wind is a stream of charged particles emitted by the Sun. By studying the composition and energy distribution of these particles, scientists were able to learn more about the Sun's composition and its influence on the space environment. The measurements also provided valuable information about the lunar surface's ability to shield astronauts from harmful radiation during future missions.

7.1.5 Cosmic Evolution and Planetary Science

The scientific discoveries made during the Apollo 11 mission had broader implications for our understanding of cosmic evolution and planetary science. By studying the Moon's composition and geological history, scientists were able to gain insights into the processes that shaped not only the Moon but also other rocky bodies in the solar system, including Earth.

The lunar samples provided a unique opportunity to study the early history of the solar system, as they preserved a record of processes that occurred billions of years ago. By comparing the lunar samples with those from other celestial bodies, scientists were able to piece together a more comprehensive picture of the solar system's formation and evolution.

The scientific discoveries and advancements made possible by the Apollo 11 mission have had a lasting impact on our understanding of the Moon, the solar system, and the universe as a whole. The knowledge gained from the analysis of lunar samples continues to shape our exploration of space and inspire future generations of scientists and engineers. The legacy of Apollo 11 extends far beyond the historic achievement of landing humans on the Moon; it has fundamentally changed our perspective on the cosmos and our place within it.

7.2 Technological Advancements

The Apollo 11 mission to the moon was not only a historic achievement for humanity, but it also marked a significant milestone in technological advancements. The successful landing on the moon was the culmination of years of research, development, and innovation. This section will explore some of the key technological advancements that were made during the Apollo 11 mission.

7.2.1 Lunar Module: The Eagle's Descent

One of the most remarkable technological advancements of the Apollo 11 mission was the development of the Lunar Module, also known as the LM. The LM, named "Eagle," was the spacecraft that carried Neil Armstrong and Buzz Aldrin to the lunar surface. It was a marvel of engineering, designed specifically for the purpose of landing on the moon.

The LM was a two-stage spacecraft, consisting of the descent stage and the ascent stage. The descent stage provided the propulsion and landing gear necessary for a safe touchdown on the lunar surface. It was equipped with a powerful rocket engine that allowed for a controlled descent and landing. The LM's landing gear, which included four landing pads, ensured a stable touchdown on the uneven lunar terrain.

The ascent stage, on the other hand, was responsible for returning the astronauts back to the Command Module, which remained in lunar orbit. It housed the crew cabin, life support systems, and the ascent engine. The ascent engine provided the necessary thrust to lift off from the moon's surface and rendezvous with the Command Module.

The development of the Lunar Module was a significant technological feat. It required the integration of various systems, including propulsion, navigation, life support, and communication. The engineers and technicians involved in its design and construction had to overcome numerous challenges, such as the need for lightweight materials, efficient fuel consumption, and reliable systems that could withstand the harsh lunar environment.

7.2.2 Command Module: The Astronauts' Home in Space

While the Lunar Module took the astronauts to the moon's surface, the Command Module served as their home during the mission. The Command Module, named "Columbia," was the primary spacecraft that carried the crew to and from the moon. It was designed to provide a safe and habitable environment for the astronauts during their journey.

The Command Module was equipped with advanced life support systems, including oxygen and temperature control, waste management, and communication equipment. It also had a navigation system that allowed the astronauts to determine their position in space accurately. The Command Module's heat shield, made of a special material called ablative material, protected the spacecraft from the intense heat generated during reentry into Earth's atmosphere.

The development of the Command Module required significant advancements in materials science, heat shield technology, and spacecraft design. The engineers had to ensure that the spacecraft could withstand the extreme temperatures and pressures experienced during reentry while keeping the astronauts safe and comfortable throughout the mission.

7.2.3 Guidance and Navigation Systems: Precision in Space

The success of the Apollo 11 mission relied heavily on the development of advanced guidance and navigation systems. These systems allowed the astronauts to navigate through space, accurately determine their position, and make precise maneuvers.

The guidance and navigation systems used during the Apollo 11 mission were a combination of onboard computers, radar systems, and star trackers. The onboard computers, known as the Apollo Guidance Computer (AGC), were state-of-the-art for their time. They were responsible for performing complex calculations, controlling the spacecraft's trajectory, and providing real-time information to the astronauts.

The radar systems played a crucial role in determining the spacecraft's distance from the moon's surface during the descent and ascent phases. They provided the astronauts with vital information to ensure a safe landing and rendezvous. The star trackers, on the other hand, allowed the astronauts to determine their position in space by comparing the positions of stars with a preloaded star catalog.

The development of these guidance and navigation systems required significant advancements in computer technology, miniaturization, and software development. The engineers had to design systems that were reliable, lightweight, and capable of operating in the harsh conditions of space.

7.2.4 Communications: Connecting Earth and Moon

Another critical technological advancement of the Apollo 11 mission was the development of a robust communication system. The astronauts needed to stay in constant contact with mission control on Earth to receive instructions, relay information, and ensure their safety.

The communication system consisted of various components, including antennas, transmitters, receivers, and tracking stations. The astronauts communicated with mission control using radio waves, which were transmitted and received by the spacecraft's antennas. The tracking stations, strategically located around the world, ensured continuous communication coverage as the spacecraft orbited the moon.

The development of the communication system required advancements in antenna design, signal processing, and transmission technology. The engineers had to overcome challenges such as signal interference, limited bandwidth, and the need for reliable communication over vast distances.

7.2.5 Materials and Technologies: Pushing the Boundaries

The Apollo 11 mission also pushed the boundaries of materials science and technology. The engineers had to develop lightweight yet robust materials that could withstand the extreme conditions of space and the moon's surface.

For example, the spacesuits worn by the astronauts were made of multiple layers of specialized materials, including fire-resistant fabrics, insulation, and airtight seals. These suits provided the astronauts with protection from the vacuum of space, extreme temperatures, micrometeoroids, and radiation.

The development of lightweight materials was also crucial for the spacecraft itself. The engineers had to find ways to reduce the weight of the spacecraft without compromising its structural integrity. This led to advancements in materials such as aluminum alloys, titanium, and composites.

Furthermore, the Apollo 11 mission also paved the way for advancements in other areas of technology, such as computer science, robotics, and medical research. The technologies developed for the mission had far-reaching implications and laid the foundation for future space exploration and scientific discoveries.

In conclusion, the Apollo 11 mission was not only a triumph of human achievement but also a testament to the remarkable technological advancements of the time. The development of the Lunar Module, Command Module, guidance and navigation systems, communication systems, and materials pushed the boundaries of what was possible. These advancements not only enabled the successful landing on the moon but also laid the foundation for future space exploration and scientific endeavors.

7.3 Inspiring Future Generations

The Apollo 11 moon landing was not only a monumental achievement for humanity, but it also served as a catalyst for inspiring future generations. The successful mission captured the imagination of people around the world and ignited a passion for space exploration that continues to this day. The bravery, determination, and ingenuity displayed by the astronauts and the entire Apollo 11 team inspired countless individuals to pursue careers in science, technology, engineering, and mathematics (STEM).

7.3.1 The Power of Inspiration

The Apollo 11 mission demonstrated the incredible power of human achievement and the boundless possibilities that exist when we push the boundaries of what is considered possible. The images and footage of Neil Armstrong taking his first steps on the lunar surface captivated the world and left an indelible mark on the collective consciousness. People from all walks of life were inspired by the audacity of the mission and the unwavering commitment of the astronauts and the entire team behind them.

7.3.2 Encouraging STEM Education

One of the most significant legacies of the Apollo 11 mission is its impact on STEM education. The moon landing sparked a renewed interest in science and technology, leading to a surge in students pursuing careers in these fields. The story of Apollo 11 served as a powerful motivator for young minds, showing them that with dedication and hard work, they too could contribute to groundbreaking discoveries and advancements.

Educational institutions and governments around the world recognized the importance of nurturing this newfound interest in STEM subjects. Scholarships, grants, and programs were established to support students pursuing degrees in science and engineering. The Apollo 11 mission also inspired the creation of science centers, museums, and educational initiatives aimed at engaging young people in hands-on learning experiences related to space exploration.

7.3.3 Diverse Representation

The Apollo 11 mission also had a profound impact on diversity and inclusion in the scientific community. The involvement of black women mathematicians, such as Katherine Johnson, Dorothy Vaughan, and Mary Jackson, highlighted the critical contributions of underrepresented groups in the field of mathematics and engineering. Their remarkable achievements shattered barriers and paved the way for future generations of women and people of color to pursue careers in STEM.

The story of these trailblazing women was brought to the forefront through the book "Hidden Figures" and the subsequent film adaptation. Their resilience, intelligence, and determination inspired countless individuals, particularly young girls and minorities, to believe in their own abilities and pursue their dreams, regardless of societal expectations or limitations.

7.3.4 Advancements in Technology

The Apollo 11 mission pushed the boundaries of technological innovation, leading to advancements that have had a lasting impact on various industries. The development of lightweight materials, miniaturized electronics, and advanced computing systems revolutionized not only space exploration but also fields such as telecommunications, medicine, and transportation.

The technologies developed for the Apollo program paved the way for the creation of satellite communication systems, which have transformed the way we communicate and access information. The miniaturization of electronics led to the development of smaller and more powerful devices, such as smartphones and laptops, that have become an integral part of our daily lives. The advancements in computing systems laid the foundation for the digital revolution and the rapid growth of the technology industry.

7.3.5 Space Exploration and Collaboration

The success of the Apollo 11 mission also fostered international collaboration in space exploration. The United States and the Soviet Union, once locked in a fierce competition known as the Space Race, began to find common ground and work together towards shared goals. This spirit of collaboration continues to shape the future of space exploration, with countries around the world joining forces to tackle the challenges of exploring the cosmos.

The International Space Station (ISS), a symbol of international cooperation, stands as a testament to the legacy of Apollo 11. Astronauts from different nations live and work together on the ISS, conducting scientific experiments and pushing the boundaries of human knowledge. The lessons learned from the Apollo program have laid the foundation for future missions to Mars and beyond, where international collaboration will be crucial for success.

7.3.6 Inspiring the Next Generation

The legacy of Apollo 11 continues to inspire the next generation of explorers, scientists, and innovators. The story of the moon landing serves as a reminder that with determination, teamwork, and a relentless pursuit of knowledge, we can achieve the seemingly impossible. The images of Neil Armstrong's first steps on the moon and the iconic words, "That's one small step for man, one giant leap for mankind," continue to resonate with people of all ages, inspiring them to dream big and reach for the stars.

Space agencies, educational institutions, and organizations around the world continue to promote the study of STEM subjects and provide opportunities for young people to engage with space exploration. From robotics competitions to space camps and mentorship programs, initiatives aimed at inspiring the next generation are thriving, ensuring that the legacy of Apollo 11 lives on in the hearts and minds of those who will shape the future.

The Apollo 11 mission was a defining moment in human history, and its impact on inspiring future generations cannot be overstated. The bravery, innovation, and collaboration displayed by the astronauts and the entire team behind the mission continue to inspire individuals to pursue their passions, push the boundaries of knowledge, and strive for greatness. The legacy of Apollo 11 serves as a reminder that when we dare to dream and work together, there is no limit to what we can achieve.

7.4 The Lasting Impact

The Apollo 11 mission to the moon was a monumental achievement for humanity. It not only marked the first time humans set foot on another celestial body but also left a lasting impact on various aspects of society. The legacy of Apollo 11 can be seen in scientific discoveries, technological advancements, and the inspiration it provided to future generations.

7.4.1 Scientific Discoveries and Lunar Samples

One of the most significant contributions of the Apollo 11 mission was the scientific knowledge gained from the lunar samples brought back to Earth. The astronauts collected approximately 47 pounds (21.5 kilograms) of moon rocks and soil during their time on the lunar surface. These samples provided scientists with invaluable insights into the moon's composition, geology, and history.

Through careful analysis, scientists discovered that the moon's surface is primarily made up of basalt, a type of volcanic rock. They also found evidence of ancient volcanic activity and impact craters that provided clues about the moon's formation and its relationship to Earth. The lunar samples helped scientists refine their understanding of the moon's age and its role in the early solar system.

Furthermore, the moon rocks contained traces of water, which challenged the previous belief that the moon was completely dry. This discovery opened up new possibilities for future lunar exploration and the potential for utilizing lunar resources.

7.4.2 Technological Advancements

The Apollo 11 mission pushed the boundaries of technology and engineering, leading to numerous advancements that have had a lasting impact on various industries. The development of the Saturn V rocket, the command module, and the lunar module required groundbreaking innovations in materials, propulsion systems, and computer technology.

The miniaturization of electronics was a crucial aspect of the Apollo program. The computers used in the Apollo spacecraft were far less powerful than today's smartphones, yet they successfully guided the astronauts to the moon and back. The advancements made in computer technology during the Apollo era laid the foundation for the digital revolution that followed.

Additionally, the Apollo program accelerated the development of materials and manufacturing techniques. Lightweight and heat-resistant materials were essential for spacecraft construction, and the techniques developed for the Apollo missions found applications in various industries, including aerospace, automotive, and construction.

7.4.3 Inspiring Future Generations

The Apollo 11 mission captured the imagination of people around the world and inspired a generation of scientists, engineers, and explorers. The sight of Neil Armstrong taking his first steps on the lunar surface was a testament to human ingenuity and the power of determination.

The success of Apollo 11 demonstrated that seemingly impossible goals could be achieved through collaboration, innovation, and perseverance. It motivated young minds to pursue careers in science, technology, engineering, and mathematics (STEM) and sparked a renewed interest in space exploration.

The impact of Apollo 11 can be seen in the subsequent missions to the moon, such as Apollo 12, 14, 15, 16, and 17. Each mission built upon the knowledge gained from its predecessors and expanded our understanding of the moon and the universe.

7.4.4 The Lasting Impact

The legacy of Apollo 11 extends far beyond the scientific and technological achievements. It symbolizes the triumph of human spirit and the pursuit of knowledge. The mission brought people together, transcending national boundaries and inspiring a sense of unity and wonder.

Apollo 11 also had a profound impact on the perception of our place in the universe. Seeing the Earth from the moon's surface provided a new perspective on our planet and its fragility. It highlighted the need for environmental stewardship and the importance of preserving our home planet for future generations.

The mission's success also had political implications. The United States' ability to land astronauts on the moon and return them safely to Earth demonstrated its technological prowess and served as a symbol of American exceptionalism during the Cold War era.

In conclusion, the Apollo 11 mission left an indelible mark on human history. Its scientific discoveries, technological advancements, and inspirational impact continue to shape our understanding of the universe and our place within it. The legacy of Apollo 11 serves as a reminder of what can be achieved when humanity dares to dream big and work together to overcome challenges.

8 The Astronauts

8.1 Neil Armstrong

Neil Armstrong, born on August 5, 1930, in Wapakoneta, Ohio, was the first man to set foot on the moon. His iconic words, "That's one small step for man, one giant leap for mankind," will forever be etched in history. Armstrong's journey to becoming an astronaut was filled with dedication, perseverance, and a deep passion for exploration.

Before joining NASA, Armstrong served as a naval aviator during the Korean War. His exceptional flying skills and ability to remain calm under pressure caught the attention of the space agency. In 1962, Armstrong became part of NASA's second group of astronauts, known as the New Nine.

As the commander of Apollo 11, Armstrong was responsible for leading the mission and piloting the Lunar Module, named Eagle, to the moon's surface. On July 20, 1969, after a four-day journey, Armstrong and Buzz Aldrin descended to the moon's surface while Michael Collins orbited above in the Command Module, Columbia.

Stepping onto the lunar surface was a moment of both triumph and uncertainty. Armstrong carefully descended the ladder of the Lunar Module, testing each step before taking his historic first step onto the moon. His actions were watched by millions of people around the world, who were captivated by the audacity and significance of the moment.

Armstrong's role extended beyond being the first man on the moon. He was also responsible for collecting samples of the lunar surface, conducting experiments, and documenting their activities. The astronauts' time on the moon was limited, and they had to make the most of every second. Armstrong and Aldrin spent approximately two and a half hours outside the Lunar Module, exploring the surface, setting up scientific instruments, and planting the American flag.

One of the most critical aspects of Armstrong's character was his humility. Despite the immense pressure and attention surrounding the mission, he remained grounded and focused on the task at hand. Armstrong's calm demeanor and exceptional piloting skills were instrumental in the successful landing and safe return of the Apollo 11 crew.

After returning from the moon, Armstrong continued to contribute to NASA and the space program. He served as the Deputy Associate Administrator for Aeronautics at NASA Headquarters and later as a professor of aerospace engineering at the University of Cincinnati. Armstrong's dedication to space exploration and his commitment to inspiring future generations of astronauts and scientists were unwavering.

Neil Armstrong's legacy extends far beyond his historic moonwalk. His achievement symbolized the triumph of human ingenuity, determination, and the relentless pursuit of knowledge. Armstrong's footsteps on the moon opened up a new era of space exploration and inspired countless individuals to reach for the stars.

Tragically, Neil Armstrong passed away on August 25, 2012, at the age of 82. However, his legacy lives on, and his name will forever be associated with one of humanity's greatest achievements. Armstrong's courage, humility, and unwavering commitment to exploration continue to inspire generations of astronauts and dreamers alike.

The first man on the moon left an indelible mark on history, reminding us that with determination and a sense of wonder, we can accomplish the seemingly impossible. Neil Armstrong's name will forever be synonymous with the Apollo 11 mission and the extraordinary feat of setting foot on another celestial body.

8.2 Buzz Aldrin

Buzz Aldrin, born Edwin Eugene Aldrin Jr. on January 20, 1930, in Glen Ridge, New Jersey, was the Lunar Module Pilot for the historic Apollo 11 mission. Aldrin's journey to becoming an astronaut was filled with determination, hard work, and a passion for exploration.

Early Life and Education

Aldrin's interest in aviation and space exploration began at a young age. His father, a military aviator, inspired him to pursue a career in the field. Aldrin attended the United States Military Academy at West Point, where he graduated third in his class with a degree in mechanical engineering in 1951.

After completing his studies at West Point, Aldrin joined the United States Air Force and flew combat missions during the Korean War. He earned the Distinguished Flying Cross for his service and later continued his education, earning a Ph.D. in astronautics from the Massachusetts Institute of Technology (MIT) in 1963.

Selection as an Astronaut

In 1963, Aldrin was selected as a member of NASA's third group of astronauts, known as the "Fourteen." This group included future space exploration pioneers such as Alan Shepard and Edgar Mitchell. Aldrin's exceptional academic background and his experience as a fighter pilot made him an ideal candidate for the Apollo program.

Training for the Moon

Aldrin, along with his fellow astronauts, underwent rigorous training to prepare for the challenges they would face during the Apollo 11 mission. This training included simulations, physical fitness exercises, and extensive study of the spacecraft systems.

One of the critical aspects of Aldrin's training was learning to operate the Lunar Module (LM), the spacecraft that would take him and Neil Armstrong to the lunar surface. Aldrin spent countless hours in simulators, practicing the intricate maneuvers required for a successful lunar landing.

The Journey to the Moon

On July 16, 1969, Aldrin, Armstrong, and Michael Collins embarked on their historic journey to the Moon aboard the Apollo 11 spacecraft. As the Lunar Module Pilot, Aldrin played a crucial role in the mission's success. He was responsible for navigating the LM and ensuring a safe landing on the lunar surface.

During the journey, Aldrin and his crewmates faced various challenges, including the physical and psychological effects of space travel. They relied on their extensive training and the support of mission control to overcome these obstacles.

Landing on the Moon

On July 20, 1969, Aldrin and Armstrong descended to the lunar surface in the Lunar Module, named "Eagle." As the LM touched down in the Sea of Tranquility, Aldrin famously radioed, "Houston, Tranquility Base here. The Eagle has landed."

Aldrin and Armstrong spent approximately two and a half hours outside the LM, conducting experiments, collecting samples, and taking photographs. Aldrin's first steps on the Moon followed Armstrong's historic "giant leap for mankind." Together, they planted the American flag and left behind a plaque commemorating their achievement.

Return to Earth and Legacy

After their successful moonwalk, Aldrin and Armstrong rejoined Collins in the Command Module, named "Columbia." They began their journey back to

Earth, carrying with them precious lunar samples and invaluable scientific data.

Upon their return, Aldrin and his crewmates were hailed as heroes and received a hero's welcome. They embarked on a worldwide tour, sharing their experiences and the significance of the Apollo 11 mission with people from all walks of life.

Buzz Aldrin's contributions to space exploration extended beyond his time as an astronaut. He continued to advocate for human space exploration and played an active role in shaping NASA's future missions. Aldrin's legacy serves as an inspiration to future generations, reminding us of the power of human ingenuity and the boundless possibilities of space exploration.

In conclusion, Buzz Aldrin's role as the Lunar Module Pilot on the Apollo 11 mission was instrumental in the success of the first moon landing. His dedication, expertise, and unwavering commitment to exploration have left an indelible mark on the history of space exploration.

8.3 Michael Collins

Michael Collins, the Command Module Pilot of Apollo 11, played a crucial role in the success of the mission. While Neil Armstrong and Buzz Aldrin descended to the lunar surface in the Lunar Module, Collins remained in orbit around the Moon in the Command Module, aptly named "Columbia." Although he did not set foot on the Moon like his fellow astronauts, Collins played a vital part in the mission's overall objective of landing humans on the lunar surface and returning them safely to Earth.

8.3.1 The Journey to the Moon

As the Command Module Pilot, Collins was responsible for piloting and navigating the Command Module during the journey to the Moon. After the Lunar Module, named "Eagle," separated from the Command Module, Collins was left alone in Columbia, orbiting the Moon at an altitude of approximately 60 miles. During this time, he had no direct communication with mission control or his fellow astronauts on the lunar surface. Collins described this period as a unique and solitary experience, where he had the opportunity to reflect on the magnitude of the mission and the significance of their endeavor.

8.3.2 The Loneliness of Space

While Armstrong and Aldrin explored the lunar surface, Collins experienced a profound sense of isolation. He was the most isolated human being in history, separated from both Earth and his fellow astronauts. Collins once said, "I am alone now, truly alone, and absolutely isolated from any known life. I am it. If a count were taken, the score would be three billion plus two over on the other side of the moon, and one plus God knows what on this side." Despite the solitude, Collins remained focused on his responsibilities and carried out his duties with precision and dedication.

8.3.3 The Importance of Collins' Role

Collins' role as the Command Module Pilot was critical to the success of the mission. While Armstrong and Aldrin explored the lunar surface, Collins was responsible for maintaining the Command Module's orbit and ensuring its safe return to Earth. He meticulously monitored the spacecraft's systems, performed necessary course corrections, and prepared for the rendezvous with the Lunar Module after its ascent from the Moon's surface. Collins' expertise and attention to detail were essential in ensuring a successful reunion with his fellow astronauts and a safe journey back to Earth.

8.3.4 Reflections on the Mission

After the mission, Collins reflected on his experience and the significance of Apollo 11. He expressed a deep sense of gratitude for being a part of such a historic event and acknowledged the teamwork and dedication of the entire Apollo 11 team. Collins recognized that while he did not walk on the Moon, his role was integral to the mission's success. He often emphasized the importance of collaboration and the collective effort required to achieve such a monumental feat.

Collins' reflections also touched on the impact of the mission on humanity as a whole. He recognized that the Apollo program had not only advanced scientific knowledge but also inspired generations of people to dream big and pursue ambitious goals. Collins believed that the spirit of exploration and the pursuit of knowledge were essential for the progress of society.

8.3.5 Life After Apollo 11

Following the Apollo 11 mission, Collins continued to contribute to the space program in various capacities. He served as the Assistant Secretary of State for Public Affairs and later as the Director of the National Air and Space Museum. Collins also authored several books, including his memoir, "Carrying the Fire," which provided a detailed account of his experiences during the Apollo 11 mission.

Throughout his life, Collins remained an advocate for space exploration and continued to inspire others with his passion for discovery. He recognized the importance of pushing the boundaries of human knowledge and believed that space exploration was a testament to humanity's curiosity and desire to explore the unknown.

Michael Collins' role as the Command Module Pilot of Apollo 11 was instrumental in the success of the mission. His dedication, expertise, and reflections on the experience continue to inspire future generations of astronauts and space enthusiasts.

8.4 The Astronauts' Experiences and Reflections

The experiences and reflections of the Apollo 11 astronauts, Neil Armstrong, Buzz Aldrin, and Michael Collins, provide a unique insight into the historic mission to the moon. Their journey was not only a physical one but also a deeply personal and emotional one. From the moment they were selected for the mission to their return to Earth, the astronauts faced numerous challenges and triumphs that shaped their perspectives on the mission and its lasting impact.

8.4.1 Neil Armstrong: The First Man on the Moon

Neil Armstrong, as the commander of Apollo 11, had the immense responsibility of being the first person to set foot on the lunar surface. His experiences and reflections offer a glimpse into the magnitude of this historic achievement. Armstrong described the moment of stepping onto the moon as "one small step for man, one giant leap for mankind." He emphasized the significance of the mission in advancing human exploration and expanding our understanding of the universe.

Armstrong's reflections also shed light on the personal and emotional aspects of the mission. He spoke of the overwhelming sense of awe and wonder he felt when looking back at Earth from the moon's surface. He described the beauty and fragility of our planet, emphasizing the need to protect and preserve it for future generations. Armstrong's experiences on the moon forever changed his perspective on life and the universe.

8.4.2 Buzz Aldrin: The Lunar Module Pilot

Buzz Aldrin, the lunar module pilot of Apollo 11, played a crucial role in the success of the mission. His experiences and reflections provide valuable insights into the challenges and triumphs of the lunar landing. Aldrin described the descent to the lunar surface as a tense and exhilarating moment. He vividly recalled the final moments before touchdown, when the Eagle lunar module's fuel was running dangerously low. The successful landing was a testament to the skill and precision of the entire Apollo 11 team.

Aldrin's reflections also touch upon the scientific and technological achievements of the mission. He spoke of the importance of collecting lunar samples and conducting experiments to further our understanding of the moon's geology and its relationship to Earth. Aldrin emphasized the significance of the mission in paving the way for future space exploration and the potential for human colonization of other celestial bodies.

8.4.3 Michael Collins: The Command Module Pilot

Michael Collins, the command module pilot of Apollo 11, had a unique perspective on the mission. While Armstrong and Aldrin descended to the lunar surface, Collins remained in orbit around the moon, providing crucial support and communication. His experiences and reflections offer a different angle on the mission and its impact.

Collins described the solitude and isolation he felt during the 48 minutes of each orbit when he was out of contact with both mission control and his fellow astronauts. He reflected on the profound sense of responsibility he felt for the success of the mission and the safety of his crewmates. Collins also spoke of the deep sense of camaraderie and trust that developed among the Apollo 11 crew during their training and mission preparations.

8.4.4 The Astronauts' Reflections on the Mission

All three astronauts shared a common reflection on the significance of the Apollo 11 mission. They recognized the immense teamwork and dedication that made the mission possible. They expressed gratitude for the support they received from mission control, the engineers, and technicians who designed and built the spacecraft, and the countless others who contributed to the success of the mission.

The astronauts also reflected on the impact of the mission on future generations. They acknowledged the inspiration it provided to young people around the world, encouraging them to pursue careers in science, technology, engineering, and mathematics. They recognized the role they played in advancing human knowledge and exploration, and the responsibility to continue pushing the boundaries of what is possible.

In their reflections, the astronauts emphasized the importance of international cooperation in space exploration. They recognized that the Apollo 11 mission was not just an American achievement but a global one. They expressed hope for continued collaboration among nations in future space missions, emphasizing the potential for scientific discoveries and advancements that benefit all of humanity.

The experiences and reflections of Neil Armstrong, Buzz Aldrin, and Michael Collins provide a profound understanding of the Apollo 11 mission. Their words capture the awe, wonder, and challenges they faced during their journey to the moon. Their reflections serve as a reminder of the human spirit's capacity for exploration, innovation, and unity in the pursuit of knowledge and understanding. The legacy of their experiences continues to inspire and shape the future of space exploration.

9 The Role of Black Women Mathematicians

9.1 Hidden Figures

The Apollo 11 mission is often remembered for the historic achievement of landing the first humans on the moon. However, behind this monumental feat, there were countless individuals who played crucial roles in making it possible. Among these unsung heroes were three remarkable black women mathematicians: Katherine Johnson, Dorothy Vaughan, and Mary Jackson. Their contributions to the Apollo program were instrumental in the success of the mission and their stories deserve to be told.

9.1.1 Katherine Johnson: A Brilliant Mind

Katherine Johnson was a brilliant mathematician whose calculations were vital to the success of the Apollo 11 mission. Born in 1918 in West Virginia, Johnson showed an early aptitude for mathematics and graduated from high school at the age of 14. She went on to earn a degree in mathematics and French from West Virginia State College.

In 1953, Johnson joined the National Advisory Committee for Aeronautics (NACA), which later became NASA. Her exceptional mathematical skills quickly caught the attention of her colleagues, and she was assigned to the Flight Research Division. Johnson's calculations were used to analyze flight paths, trajectories, and launch windows for various missions.

During the Apollo program, Johnson's calculations were crucial in determining the trajectory for the lunar module to land safely on the moon. Her work involved complex calculations to account for the gravitational forces of the Earth, moon, and other celestial bodies. Johnson's accuracy and attention to detail were unparalleled, and her calculations were relied upon by the astronauts and mission control.

9.1.2 Dorothy Vaughan: A Trailblazer in Computing

Dorothy Vaughan was another remarkable mathematician who made significant contributions to the Apollo program. Born in 1910 in Kansas, Vaughan showed an early interest in mathematics and pursued a degree in mathematics and physics from Wilberforce University.

In 1943, Vaughan joined the NACA's segregated West Area Computing Unit, where she became the first black supervisor. She was responsible for leading a team of black women mathematicians known as the "West Computers." Vaughan's team played a crucial role in performing complex calculations for various projects, including the Apollo program.

Vaughan's expertise in computer programming was particularly valuable during the transition from human computers to electronic computers. She taught herself programming languages and became proficient in Fortran, a programming language used by NASA. Vaughan's knowledge and leadership were instrumental in ensuring a smooth transition to electronic computing, which greatly enhanced the efficiency and accuracy of calculations for the Apollo missions.

9.1.3 Mary Jackson: Breaking Barriers

Mary Jackson, born in 1921 in Virginia, was a talented mathematician and engineer who made significant contributions to the Apollo program. After graduating from Hampton Institute with degrees in mathematics and physical science, Jackson joined the NACA in 1951.

Jackson initially worked as a research mathematician, but her passion for engineering led her to pursue a career change. Despite facing racial and gender barriers, Jackson became NASA's first black female engineer in 1958. Her expertise in aerodynamics and wind tunnel testing proved invaluable in the development of spacecraft and astronaut training.

Jackson's determination and perseverance paved the way for future generations of black women in STEM fields. She actively promoted equal opportunities for women and minorities in the aerospace industry and served as a mentor to many aspiring engineers.

9.1.4 The Hidden Figures' Impact

The contributions of Katherine Johnson, Dorothy Vaughan, and Mary Jackson were not only critical to the success of the Apollo 11 mission but also had a lasting impact on the advancement of civil rights and gender equality. Their achievements shattered stereotypes and opened doors for future generations of black women in mathematics, engineering, and other STEM fields.

Despite facing discrimination and prejudice, these remarkable women persevered and excelled in their respective fields. Their dedication, intelligence, and unwavering commitment to their work paved the way for a more inclusive and diverse space program.

In 2015, their stories were brought to the forefront with the publication of the book "Hidden Figures" by Margot Lee Shetterly, which was later adapted into a critically acclaimed film. The book and film shed light on the often-overlooked contributions of these extraordinary women and brought their stories to a wider audience.

Today, the legacy of Katherine Johnson, Dorothy Vaughan, and Mary Jackson lives on as an inspiration to aspiring mathematicians, engineers, and scientists around the world. Their remarkable achievements serve as a reminder that talent knows no boundaries and that diversity and inclusion are essential for progress and innovation.

9.2 Their Contributions to the Apollo Program

The Apollo 11 mission to the moon was a monumental achievement for humanity, and it would not have been possible without the contributions of countless individuals. Among these unsung heroes were the black women mathematicians who played a crucial role in the success of the Apollo program. Katherine Johnson, Dorothy Vaughan, and Mary Jackson, also known as the "Hidden Figures," made significant contributions to the mission through their exceptional mathematical skills and determination.

9.2.1 Katherine Johnson: Calculating Trajectories

Katherine Johnson, a brilliant mathematician, played a vital role in the Apollo program by calculating the trajectories for the spacecraft. Her calculations were essential for ensuring the safe and accurate navigation of the Apollo missions. Johnson's expertise in celestial mechanics and her ability to perform complex calculations by hand were invaluable to the success of the program.

Johnson's work involved analyzing data from various sources, including radar and telemetry, to determine the spacecraft's position and velocity. Her calculations helped the astronauts navigate through space, ensuring that they reached their intended destinations with precision. Johnson's contributions were particularly critical during the Apollo 11 mission, as she helped calculate the trajectory for the lunar module's descent to the moon's surface.

9.2.2 Dorothy Vaughan: Human Computer Supervisor

Dorothy Vaughan was another remarkable mathematician who made significant contributions to the Apollo program. As a supervisor of the West Area Computing Unit at NASA's Langley Research Center, Vaughan played a crucial role in managing a team of black women mathematicians who were responsible for performing complex calculations.

Vaughan's leadership and expertise were instrumental in transitioning from manual calculations to electronic computing. She recognized the importance of computer programming and taught herself and her team the programming language FORTRAN. This knowledge allowed them to program the newly installed IBM 7090 computer, which greatly enhanced the efficiency and accuracy of calculations for the Apollo missions.

Vaughan's contributions extended beyond her technical skills. She was a trailblazer for racial equality, advocating for equal opportunities for black women in the workplace. Her determination and leadership paved the way for future generations of black women mathematicians and engineers.

9.2.3 Mary Jackson: Aerospace Engineer

Mary Jackson, an aerospace engineer, made significant contributions to the Apollo program through her expertise in aerodynamics and engineering. Jackson started her career as a mathematician at NASA and later became the first black female engineer at the agency.

Jackson's work involved conducting research and performing calculations to improve the design and performance of aircraft and spacecraft. Her contributions to the Apollo program included analyzing data from wind tunnel tests and developing innovative solutions to engineering challenges.

Jackson's expertise was particularly crucial during the development of the Apollo command module. She played a vital role in ensuring the spacecraft's aerodynamic stability and safety during reentry into Earth's atmosphere. Her contributions helped make the Apollo missions a success and paved the way for future advancements in aerospace engineering.

9.2.4 Collaborative Efforts and Impact

The contributions of Katherine Johnson, Dorothy Vaughan, and Mary Jackson were not only significant individually but also as a collective effort. Their collaboration and dedication to their work were instrumental in the success of the Apollo program.

Their achievements also had a profound impact beyond the Apollo missions. Their groundbreaking work challenged the prevailing societal norms and broke down barriers for women and minorities in the fields of science, technology, engineering, and mathematics (STEM). Their accomplishments inspired future generations to pursue careers in these fields and continue pushing the boundaries of human knowledge and exploration.

The recognition of the "Hidden Figures" came decades after their contributions, but their legacy continues to inspire and empower individuals from all backgrounds. Their story serves as a reminder of the importance of diversity and inclusion in scientific and technological advancements.

The Apollo 11 mission would not have been possible without the remarkable contributions of Katherine Johnson, Dorothy Vaughan, and Mary Jackson. Their exceptional mathematical skills, determination, and perseverance played a vital role in the success of the mission and the broader Apollo program. Their story is a testament to the power of human ingenuity and the impact that individuals can have when given the opportunity to excel.

9.3 Overcoming Challenges and Breaking Barriers

The story of Apollo 11 and the moon landing is not just about the astronauts and the engineers who made it possible. It is also a story of the remarkable contributions made by black women mathematicians, who overcame numerous challenges and broke down barriers to play a critical role in the success of the mission.

9.3.1 The Struggles of Black Women Mathematicians

In the 1960s, racial and gender discrimination were pervasive in American society. Black women faced significant obstacles in pursuing careers in mathematics and science. Despite their exceptional talents and qualifications, they were often overlooked and marginalized. However, a few remarkable women managed to overcome these challenges and make their mark in the field.

9.3.2 Katherine Johnson: A Trailblazer in Mathematics

One of the most prominent figures among these women was Katherine Johnson. Born in 1918, Johnson showed an early aptitude for mathematics. She graduated summa cum laude with degrees in mathematics and French from West Virginia State College. Despite her qualifications, she faced numerous barriers when seeking employment as a mathematician.

In 1953, Johnson joined the National Advisory Committee for Aeronautics (NACA), which later became NASA. She worked as a "human computer," performing complex calculations for various projects. Her exceptional skills and attention to detail quickly earned her a reputation as one of the most talented mathematicians at NASA.

9.3.3 Dorothy Vaughan: A Leader and Mentor

Dorothy Vaughan was another remarkable woman who played a crucial role in the success of the Apollo 11 mission. Like Johnson, Vaughan faced significant challenges in her career due to racial and gender discrimination. However, her determination and perseverance led her to become the first African American supervisor at NACA.

Vaughan's leadership skills and technical expertise were instrumental in the success of the early space missions. She was responsible for leading a team of black women mathematicians known as the "West Area Computers." Under her guidance, these women made significant contributions to the space program, including the Apollo missions.

9.3.4 Mary Jackson: Breaking Barriers in Engineering

Mary Jackson, another brilliant mathematician, faced similar challenges in her career. Despite her exceptional skills, she was initially denied the opportunity to become an engineer due to segregation and discrimination. However, Jackson's determination and perseverance led her to fight for her rights and break down barriers.

In 1958, Jackson became NASA's first black female engineer. Her expertise in aerodynamics was invaluable in the development of the space program. She worked on various projects, including the Apollo missions, where her calculations and insights played a crucial role in ensuring the safety and success of the astronauts.

9.3.5 Overcoming Challenges and Making History

These remarkable women faced numerous challenges and overcame significant obstacles to contribute to the success of the Apollo 11 mission. They had to navigate a society that often dismissed their talents and capabilities. However, their dedication, intelligence, and perseverance allowed them to break down barriers and make history.

Their contributions were not limited to their technical skills alone. They also served as role models and mentors for future generations of black women mathematicians and scientists. Their achievements inspired countless others to pursue careers in STEM fields and showed that talent knows no boundaries.

9.3.6 Recognition and Legacy

It is essential to acknowledge that the contributions of these women were not widely recognized at the time. Their stories remained largely untold until the publication of Margot Lee Shetterly's book, "Hidden Figures," which brought their achievements to the forefront. The subsequent film adaptation further shed light on their remarkable journey.

In recent years, there has been a growing recognition of the vital role played by black women mathematicians in the success of the Apollo program. Their contributions have been celebrated, and they have received numerous awards and honors for their groundbreaking work.

The legacy of Katherine Johnson, Dorothy Vaughan, Mary Jackson, and other black women mathematicians lives on in the countless individuals they have inspired. Their story serves as a reminder that talent and determination can overcome even the most significant barriers. Their contributions to the Apollo 11 mission and the field of mathematics will forever be remembered as a testament to the power of perseverance and the importance of diversity in scientific endeavors.

9.4 Recognition and Legacy

The contributions of black women mathematicians to the Apollo 11 mission were invaluable and their legacy continues to inspire generations. Katherine Johnson, Dorothy Vaughan, and Mary Jackson, also known as the "Hidden Figures," played a crucial role in the success of the Apollo program. Their remarkable achievements and perseverance in the face of adversity have left an indelible mark on history.

9.4.1 Recognition and Appreciation

Despite facing discrimination and prejudice, Katherine Johnson, Dorothy Vaughan, and Mary Jackson made significant contributions to the Apollo 11 mission. Their mathematical expertise and problem-solving skills were instrumental in the success of the mission. Their work involved calculating trajectories, analyzing data, and ensuring the accuracy of the spacecraft's navigation systems.

Although their contributions were essential, the recognition they received at the time was limited. The prevailing societal norms and racial biases of the era meant that their achievements often went unnoticed or were overshadowed. However, their colleagues and those who worked closely with them recognized their immense talent and dedication.

9.4.2 Post-Apollo 11 Recognition

In the years following the Apollo 11 mission, the contributions of Katherine Johnson, Dorothy Vaughan, and Mary Jackson began to receive the recognition they deserved. Their stories were brought to light through books, documentaries, and interviews, allowing the world to learn about their remarkable achievements.

In 2016, Margot Lee Shetterly's book "Hidden Figures" was published, shedding light on the lives and work of these extraordinary women. The book was later adapted into a critically acclaimed film, further amplifying their stories and inspiring a new generation of scientists and mathematicians.

9.4.3 Awards and Honors

As their contributions became more widely known, Katherine Johnson, Dorothy Vaughan, and Mary Jackson received numerous awards and honors for their groundbreaking work. In 2015, Katherine Johnson was awarded the Presidential Medal of Freedom, the highest civilian honor in the United States, for her pioneering contributions to space exploration.

Dorothy Vaughan and Mary Jackson were posthumously awarded the Congressional Gold Medal in 2019, recognizing their significant contributions to the success of the Apollo program. These honors not only acknowledged their individual achievements but also symbolized the recognition of the countless other black women who made important contributions to the space program.

9.4.4 Inspiring Future Generations

The legacy of Katherine Johnson, Dorothy Vaughan, and Mary Jackson extends far beyond their individual accomplishments. Their stories have become a source of inspiration for countless individuals, particularly young women and people of color, who aspire to pursue careers in science, technology, engineering, and mathematics (STEM).

Their determination to overcome barriers and excel in their fields serves as a powerful reminder that talent and intelligence are not limited by race or gender. Their stories have encouraged a more inclusive and diverse approach to STEM fields, fostering an environment where everyone has the opportunity to contribute and succeed.

9.4.5 Breaking Barriers and Opening Doors

The contributions of black women mathematicians to the Apollo 11 mission shattered stereotypes and opened doors for future generations. Their achievements challenged the prevailing notions of who could excel in scientific and technical fields, paving the way for greater diversity and inclusion in the aerospace industry.

Their legacy continues to inspire initiatives aimed at promoting diversity and equality in STEM fields. Organizations and institutions have recognized the importance of providing opportunities and support for underrepresented groups, ensuring that the next generation of scientists and mathematicians reflects the diversity of society.

9.4.6 The Enduring Legacy

The legacy of Katherine Johnson, Dorothy Vaughan, and Mary Jackson is one of resilience, determination, and excellence. Their contributions to the Apollo 11 mission and the broader field of space exploration have left an indelible mark on history. Their stories serve as a reminder that the pursuit of knowledge and exploration knows no boundaries.

Their recognition and legacy have sparked a renewed interest in the untold stories of other hidden figures who played pivotal roles in shaping history. Their achievements have become a symbol of the triumph of human potential and the power of diversity in driving innovation and progress.

As we reflect on the historic achievement of the Apollo 11 mission, it is essential to acknowledge and celebrate the contributions of all those involved, including the remarkable black women mathematicians who defied the odds and helped humanity take one giant leap towards the moon. Their recognition and legacy will continue to inspire and shape the future of space exploration for generations to come.

10 Trials and Challenges

10.1 Apollo 1

The Apollo 1 mission was intended to be the first manned mission of the Apollo program, with the goal of testing the Command and Service Module (CSM) in low Earth orbit. However, tragedy struck on January 27, 1967, during a pre-launch test when a fire broke out inside the Apollo 1 spacecraft, resulting in the deaths of all three astronauts on board: Virgil "Gus" Grissom, Edward H. White II, and Roger B. Chaffee.

The Apollo 1 mission was originally scheduled for launch on February 21, 1967, but the tragic accident led to a thorough investigation and a reassessment of the spacecraft's design and safety protocols. The investigation revealed several design flaws and safety issues that contributed to the fire and subsequent loss of the crew.

The fire started in the pure oxygen environment of the spacecraft's cabin, which was pressurized to simulate the conditions of space. The highly flammable materials used in the spacecraft's construction, such as nylon netting and Velcro, allowed the fire to spread rapidly. Additionally, the crew was unable to quickly open the hatch due to a design flaw, further exacerbating the situation.

The loss of the Apollo 1 crew was a devastating blow to NASA and the entire nation. It was a stark reminder of the risks involved in space exploration and the need for rigorous safety measures. The tragedy prompted a complete overhaul of the Apollo spacecraft's design and safety protocols, ensuring that such a disaster would never happen again.

The investigation into the Apollo 1 fire led to significant changes in the spacecraft's design. The pure oxygen environment was replaced with a mixture of oxygen and nitrogen, reducing the risk of fire. The flammable materials were replaced with fire-resistant alternatives, and the hatch design was modified to allow for quick and easy opening in case of an emergency.

The lessons learned from the Apollo 1 tragedy also had a profound impact on the training and preparation of future astronauts. Safety became a top priority, and extensive measures were put in place to ensure the well-being of the crew during all phases of the mission. Emergency procedures were refined, and astronauts underwent rigorous training in fire safety and emergency escape.

The loss of the Apollo 1 crew also had a lasting impact on the morale of NASA and the American public. It was a stark reminder of the risks involved in space exploration and the sacrifices made by the astronauts who ventured into the unknown. However, it also served as a catalyst for change and a renewed commitment to the goal of reaching the moon.

The tragedy of Apollo 1 also highlighted the importance of the work being done by the black women mathematicians at NASA. Katherine Johnson, Dorothy Vaughan, and Mary Jackson, who were featured in the book and movie "Hidden Figures," played a critical role in the Apollo program. Their calculations and expertise were instrumental in the success of the missions, including Apollo 11.

In the aftermath of the Apollo 1 fire, NASA made a concerted effort to improve diversity and inclusion within the agency. The contributions of black women mathematicians like Johnson, Vaughan, and Jackson were finally recognized and celebrated. Their groundbreaking work paved the way for future generations of women and minorities in the field of science and engineering.

The Apollo 1 tragedy was a somber reminder of the risks and challenges inherent in space exploration. It was a pivotal moment in the history of the Apollo program, leading to significant changes in spacecraft design, safety protocols, and astronaut training. The loss of the Apollo 1 crew will forever be remembered as a sacrifice made in the pursuit of knowledge and the exploration of the unknown.

10.2 Learning from Mistakes

The Apollo program was an ambitious endeavor that aimed to put humans on the moon for the first time in history. However, it was not without its fair share of trials and challenges. In this section, we will explore some of the mistakes made during the program and the valuable lessons learned from them.

10.2.1 Apollo 1: The Tragedy

One of the most significant setbacks in the Apollo program occurred on January 27, 1967, during a pre-launch test for the Apollo 1 mission. Astronauts Gus Grissom, Ed White, and Roger Chaffee were inside the command module when a fire broke out, resulting in the tragic loss of all three lives. This devastating event served as a wake-up call for NASA and the entire space community.

The investigation into the Apollo 1 fire revealed several critical design and safety flaws in the command module. The highly pressurized pure oxygen environment inside the spacecraft, combined with flammable materials, created a hazardous situation. Additionally, the hatch design made it difficult for the astronauts to escape in an emergency.

10.2.2 Learning from Mistakes

The Apollo 1 tragedy prompted NASA to reevaluate its approach to spacecraft design, safety protocols, and astronaut training. The agency implemented a series of changes to prevent similar accidents in the future.

First and foremost, NASA modified the command module's design to improve safety. The pure oxygen atmosphere was replaced with a mixture of oxygen and nitrogen, reducing the risk of fire. The hatch design was also redesigned to allow for quick and easy opening in case of an emergency.

Furthermore, NASA established stricter safety protocols and procedures. The agency implemented rigorous testing and quality control measures to ensure the reliability and safety of the spacecraft. Astronauts underwent extensive training in emergency procedures and were equipped with the knowledge and skills to handle critical situations.

10.2.3 Apollo 8: The First Lunar Orbit

After the tragedy of Apollo 1, NASA was determined to get the Apollo program back on track. The next mission, Apollo 8, would be a crucial step towards achieving the goal of landing humans on the moon. On December 21, 1968, Apollo 8 launched with astronauts Frank Borman, James Lovell, and William Anders on board.

Apollo 8 was the first mission to orbit the moon, providing valuable insights into the lunar environment and paving the way for future lunar landings. The mission was not without its challenges, though. The crew faced the risk of radiation exposure from the Van Allen belts and the uncertainty of the lunar module's performance.

Despite these challenges, Apollo 8 was a resounding success. The crew successfully entered lunar orbit, captured stunning photographs of the moon's surface, and delivered a live television broadcast from space that captivated the world. The mission demonstrated NASA's ability to overcome obstacles and marked a significant milestone in the Apollo program.

10.2.4 Apollo 10: The Dress Rehearsal

Before attempting a lunar landing, NASA planned a dress rehearsal mission to test all the necessary procedures and systems. Apollo 10, launched on May 18, 1969, with astronauts Thomas Stafford, John Young, and Eugene Cernan on board, served as the final test before the historic Apollo 11 mission.

Apollo 10 aimed to simulate all aspects of a lunar landing, except for the actual touchdown. The crew successfully performed a lunar orbit insertion, tested the lunar module's descent and ascent stages, and conducted a rendezvous and docking maneuver in lunar orbit. The mission provided valuable data on the lunar module's performance and allowed NASA to fine-tune the procedures for the upcoming Apollo 11 mission.

Despite the success of Apollo 10, the mission also highlighted some areas for improvement. The crew encountered difficulties during the lunar module's ascent stage, experiencing a roll oscillation that could have jeopardized a safe return to the command module. This issue was addressed and resolved before the Apollo 11 mission.

Conclusion

The Apollo program faced numerous trials and challenges on its path to the moon. The tragic loss of the Apollo 1 crew served as a stark reminder of the risks involved in space exploration. However, NASA's ability to learn from its mistakes and make necessary improvements ensured the success of subsequent missions.

The lessons learned from the Apollo 1 tragedy, along with the experiences gained from missions like Apollo 8 and Apollo 10, played a crucial role in the ultimate success of Apollo 11. The mistakes made along the way served as valuable stepping stones towards achieving the historic moon landing and left a lasting legacy for future space exploration endeavors.

10.3 Apollo 8

Apollo 8 was a pivotal mission in the Apollo program that paved the way for the historic moon landing of Apollo 11. It was the first mission to send humans to orbit the Moon and marked a significant milestone in space exploration. The crew of Apollo 8 consisted of Commander Frank Borman, Command Module Pilot James Lovell, and Lunar Module Pilot William Anders. This daring mission not only tested the capabilities of the spacecraft but also provided valuable insights and experience for future lunar missions.

10.3.1 The Decision to Orbit the Moon

After the tragic Apollo 1 fire, NASA faced numerous challenges and setbacks. The loss of three astronauts highlighted the need for thorough testing and safety measures. As the Apollo program regrouped, the decision was made to change the mission plan for Apollo 8. Originally intended as a test flight in Earth orbit, it was now proposed that Apollo 8 would become the first mission to orbit the Moon.

10.3.2 Preparations and Training

The crew of Apollo 8 underwent extensive training to prepare for their groundbreaking mission. They studied the lunar surface, practiced navigation techniques, and familiarized themselves with the spacecraft systems. Simulations and mock missions were conducted to simulate the conditions they would face in space. The crew also received training on the Lunar Module, even though it would not be used on this particular mission.

10.3.3 The Journey to the Moon

On December 21, 1968, Apollo 8 lifted off from Kennedy Space Center atop a Saturn V rocket. The spacecraft successfully entered Earth orbit and then performed a critical Trans Lunar Injection (TLI) burn to set a course for the Moon. As they traveled towards their destination, the crew captured awe-inspiring images of Earth, including the iconic "Earthrise" photograph.

10.3.4 Lunar Orbit and Christmas Message

After a journey of approximately three days, Apollo 8 entered lunar orbit on December 24, 1968. This marked the first time humans had ever orbited another celestial body. As they circled the Moon, the crew conducted experiments, took photographs, and gathered valuable data about the lunar surface. On Christmas Eve, the crew delivered a live television broadcast, during which they read from the Book of Genesis, sharing a message of peace and unity with the world.

10.3.5 Return to Earth

Following their successful lunar orbit, Apollo 8 began its journey back to Earth. The crew performed a critical Trans Earth Injection (TEI) burn to set a trajectory for reentry. The Command Module reentered Earth's atmosphere and splashed down safely in the Pacific Ocean on December 27, 1968. The crew was recovered by the USS Yorktown and greeted as heroes upon their return.

10.3.6 Legacy and Impact

Apollo 8 was a resounding success and a testament to human ingenuity and determination. It demonstrated that humans could travel to the Moon and return safely, setting the stage for the subsequent Apollo missions. The mission provided valuable data on the lunar environment and helped refine the techniques and procedures necessary for future lunar landings.

The crew of Apollo 8, along with the dedicated team in mission control, played a crucial role in the success of the mission. Their expertise, courage, and unwavering commitment to the goal of space exploration propelled humanity to new heights. The mission also highlighted the importance of teamwork and collaboration, as the crew and ground personnel worked together seamlessly to overcome challenges and achieve their objectives.

The success of Apollo 8 had a profound impact on the public's perception of space exploration. It captured the imagination of people around the world and inspired a sense of wonder and possibility. The mission demonstrated the power of human achievement and the potential for further exploration beyond Earth.

In conclusion, Apollo 8 was a groundbreaking mission that set the stage for the historic moon landing of Apollo 11. It showcased the capabilities of the spacecraft, tested the skills of the crew, and provided valuable insights for future lunar missions. The success of Apollo 8 not only advanced the goals of the Apollo program but also ignited a sense of wonder and possibility in the hearts and minds of people worldwide.

10.4 Apollo 10

Apollo 10, often referred to as the "dress rehearsal" for the historic Apollo 11 mission, played a crucial role in paving the way for the successful moon landing. This mission, which took place in May 1969, aimed to test all aspects of the lunar module and the command module in lunar orbit, including the critical descent and ascent stages. Led by Commander Thomas P. Stafford, Command Module Pilot John W. Young, and Lunar Module Pilot Eugene A. Cernan, Apollo 10 was a pivotal step in ensuring the safety and success of the upcoming Apollo 11 mission.

10.4.1 Preparing for Lunar Orbit

Apollo 10 began its journey on May 18, 1969, with the launch of the Saturn V rocket from Kennedy Space Center in Florida. The crew, consisting of experienced astronauts, had undergone extensive training and simulations to prepare for the challenges they would face during the mission. As the spacecraft hurtled towards the Moon, the crew's excitement and anticipation grew, knowing that they were on the cusp of making history.

10.4.2 Lunar Module Separation

Upon reaching lunar orbit, the crew prepared for a critical maneuver: the separation of the lunar module, nicknamed "Snoopy," from the command module, known as "Charlie Brown." This separation was a crucial step in the mission, as it would allow the crew to test the lunar module's descent and ascent capabilities without actually landing on the Moon's surface.

Commander Stafford and Lunar Module Pilot Cernan boarded Snoopy, while Command Module Pilot Young remained in Charlie Brown. The separation was successful, and Snoopy descended to within just 50,000 feet of the lunar surface. This marked the closest any human had ever been to the Moon at that time.

10.4.3 Testing the Lunar Module

During the descent, Stafford and Cernan meticulously tested the lunar module's systems, ensuring that everything was functioning as expected. They evaluated the navigation, communication, and propulsion systems, as well as the module's ability to rendezvous and dock with the command module. These tests were crucial in identifying any potential issues that could arise during the actual lunar landing.

As Snoopy descended towards the lunar surface, the crew encountered a few unexpected challenges. They experienced a phenomenon known as "slosh," where the fuel in the lunar module's tanks caused the spacecraft to oscillate. This was a valuable lesson for future missions, as it highlighted the need for improved fuel management systems.

10.4.4 The Return Journey

After completing their tests and gathering valuable data, Stafford and Cernan rejoined Young in the command module. The crew then began their journey back to Earth, leaving the lunar module behind in lunar orbit. As they made their way home, they reflected on the significance of their mission and the role they played in preparing for the historic Apollo 11 landing.

10.4.5 Lessons Learned and Legacy

Apollo 10 provided invaluable insights and lessons that directly influenced the success of the Apollo 11 mission. The crew's experiences and the data they collected during their dress rehearsal were instrumental in refining the lunar landing procedures and ensuring the safety of the astronauts who would follow in their footsteps.

The mission also highlighted the incredible teamwork and dedication of the engineers, technicians, and support staff who worked tirelessly behind the scenes. Their expertise and attention to detail were critical in overcoming the challenges faced during the mission and ultimately achieving the goal of landing humans on the Moon.

Apollo 10's legacy extends beyond its immediate contributions to the Apollo program. It served as a testament to human ingenuity, pushing the boundaries of what was thought possible and inspiring future generations to dream big. The success of Apollo 10 set the stage for the historic Apollo 11 mission, which would forever change our understanding of the universe and our place within it.

In conclusion, Apollo 10 played a pivotal role in the Apollo program, serving as the dress rehearsal for the first moon landing. The mission tested the lunar module's capabilities, identified potential issues, and provided valuable data that directly influenced the success of Apollo 11. The crew's dedication and the support of the entire team behind the scenes ensured the safety and success of the mission. Apollo 10's legacy lives on as a testament to human achievement and the enduring spirit of exploration.

11 Behind the Scenes

11.1 The Engineers and Technicians

Behind every successful mission, there is a team of dedicated engineers and technicians who work tirelessly to design, build, and test the spacecraft and its components. The Apollo 11 mission was no exception. This section will delve into the crucial role played by these unsung heroes in making the historic moon landing possible.

11.1.1 The Design Process

The engineers and technicians involved in the Apollo 11 mission faced an enormous challenge: designing a spacecraft capable of safely transporting astronauts to the moon and back. This required a meticulous design process that involved countless hours of research, testing, and collaboration.

The design of the Apollo spacecraft was a collaborative effort between NASA and various aerospace companies, including North American Aviation (NAA), which was responsible for the command and service modules, and Grumman Aircraft Engineering Corporation, which designed the lunar module. These companies assembled teams of engineers and technicians who worked together to bring President Kennedy's vision to life.

The design process began with extensive research and development. Engineers studied previous space missions, analyzed data, and conducted simulations to understand the unique challenges of a lunar landing. They had to consider factors such as weight limitations, fuel efficiency, and the harsh conditions of space.

Once the initial design concepts were established, engineers and technicians began the detailed work of creating blueprints and specifications for each component of the spacecraft. This involved a meticulous attention to detail, as even the smallest error could have catastrophic consequences.

11.1.2 Building the Spacecraft

With the design finalized, the engineers and technicians turned their attention to building the spacecraft. This involved a complex manufacturing process that required precision and expertise.

The construction of the command and service modules took place at the NAA's facilities in California. Skilled technicians meticulously assembled each component, ensuring that they met the strict quality standards set by NASA. The process involved welding, machining, and testing to ensure that the modules could withstand the extreme conditions of space.

Meanwhile, the lunar module was being built by Grumman Aircraft Engineering Corporation in New York. This unique spacecraft had to be lightweight yet sturdy enough to withstand the lunar landing. The engineers and technicians at Grumman worked tirelessly to construct a module that would serve as the astronauts' home on the moon's surface.

Throughout the construction process, quality control measures were implemented to ensure that every component met the highest standards. Inspections, tests, and simulations were conducted to identify and rectify any potential issues before the spacecraft was deemed flight-ready.

11.1.3 Testing and Quality Control

Before the Apollo 11 mission could proceed, the spacecraft had to undergo rigorous testing to ensure its reliability and safety. This involved a series of tests that simulated the conditions the spacecraft would encounter during launch, in space, and during reentry.

One of the most critical tests was the "pad abort test," which simulated an emergency escape scenario during launch. This test evaluated the spacecraft's ability to safely separate from the rocket in the event of an anomaly. The engineers and technicians meticulously analyzed the data from these tests, making necessary adjustments to improve the spacecraft's performance.

Another crucial test was the "lunar module landing simulation," which replicated the conditions the astronauts would face during the lunar landing. This test allowed the engineers and technicians to fine-tune the module's descent and landing procedures, ensuring a safe touchdown on the moon's surface.

Throughout the testing phase, the engineers and technicians worked closely with the astronauts to gather feedback and make necessary modifications. This collaborative approach ensured that the spacecraft met the specific needs of the mission and the astronauts.

11.1.4 Supporting the Mission

The engineers and technicians involved in the Apollo 11 mission did not stop at designing, building, and testing the spacecraft. They also played a crucial role in supporting the mission from the ground.

During the mission, a team of engineers and technicians manned the mission control center at NASA's Johnson Space Center in Houston, Texas. These individuals monitored the spacecraft's systems, communicated with the astronauts, and made critical decisions in real-time to ensure the mission's success.

The engineers and technicians in mission control were responsible for analyzing data, troubleshooting issues, and providing guidance to the astronauts. Their expertise and quick thinking were instrumental in overcoming challenges and ensuring the safe return of the crew.

In addition to mission control, a vast network of engineers and technicians supported the mission behind the scenes. From tracking the spacecraft's trajectory to maintaining communication systems, these individuals worked tirelessly to ensure that every aspect of the mission ran smoothly.

The engineers and technicians involved in the Apollo 11 mission were the unsung heroes who made the moon landing possible. Their dedication, expertise, and attention to detail were instrumental in designing, building, and testing the spacecraft. Their support from mission control and behind the scenes ensured the success of the mission. Without their tireless efforts, the historic achievement of Apollo 11 would not have been possible.

11.2 Designing and Building the Spacecraft

The success of the Apollo 11 mission relied heavily on the meticulous design and construction of the spacecraft. The engineers and technicians involved in this process faced numerous challenges and had to overcome various obstacles to ensure the safety and functionality of the spacecraft.

11.2.1 The Lunar Module: Eagle

One of the most critical components of the Apollo 11 mission was the Lunar Module, also known as the Eagle. This spacecraft was specifically designed to land on the lunar surface and allow the astronauts to explore and conduct experiments. The development of the Lunar Module required innovative engineering solutions to overcome the challenges of landing on the moon.

The design of the Lunar Module had to take into account the harsh lunar environment, including the lack of atmosphere and the uneven surface. The engineers had to ensure that the spacecraft could withstand the landing impact and provide a stable platform for the astronauts. The structure of the Lunar Module was made from lightweight materials to reduce its weight and allow for efficient fuel consumption during descent and ascent.

The construction of the Lunar Module involved a collaborative effort between various contractors and NASA's own engineers. Grumman Aircraft Engineering Corporation was responsible for the overall design and assembly of the Lunar Module. The spacecraft consisted of two main sections: the descent stage and the ascent stage.

The descent stage housed the landing gear, engines, and fuel tanks. It provided the necessary thrust and control for a controlled descent to the lunar surface. The descent engine, known as the Descent Propulsion System (DPS), was responsible for slowing down the spacecraft and safely landing it on the moon.

The ascent stage, on the other hand, contained the crew cabin, life support systems, and the ascent engine. The ascent engine, called the Ascent Propulsion System (APS), would later propel the astronauts back to the Command Module for their return journey to Earth.

11.2.2 The Command and Service Module: Columbia

While the Lunar Module was designed for lunar landing and exploration, the Command and Service Module (CSM) played a crucial role in the overall mission. The CSM, named Columbia, served as the primary spacecraft for the astronauts during their journey to the moon, lunar orbit, and the return to Earth.

The CSM consisted of two main components: the Command Module (CM) and the Service Module (SM). The Command Module housed the crew during the entire mission and provided the necessary life support systems, communication equipment, and controls. It was the only part of the spacecraft that would return to Earth.

The Service Module, located beneath the Command Module, contained the propulsion system, fuel cells, and other essential systems required for the mission. It provided the necessary power and propulsion for the spacecraft, including the critical Trans-Earth Injection (TEI) burn that would send the astronauts back on their trajectory towards Earth.

The design and construction of the Command and Service Module were carried out by North American Aviation (later known as North American Rockwell). The engineers faced numerous challenges in developing a spacecraft that could withstand the harsh conditions of space and ensure the safety of the astronauts.

The Command Module was designed to have a conical shape to minimize aerodynamic drag during reentry into Earth's atmosphere. It was equipped with a heat shield made of ablative material, which would burn away during reentry, dissipating the intense heat generated by atmospheric friction.

The Service Module, located beneath the Command Module, housed the main engine, propellant tanks, and other critical systems. It provided the necessary propulsion for various mission maneuvers, including the crucial Trans-Lunar Injection (TLI) burn that propelled the spacecraft towards the moon.

11.2.3 Collaboration and Quality Control

The design and construction of the Apollo spacecraft involved a collaborative effort between NASA, contractors, and various subcontractors. The process required meticulous attention to detail and rigorous quality control measures to ensure the reliability and safety of the spacecraft.

NASA established a comprehensive system of checks and balances to monitor the design and construction process. The agency conducted regular reviews and inspections to ensure that the spacecraft met the required specifications and standards. This involved extensive testing of individual components, subsystems, and the complete spacecraft.

The contractors and subcontractors involved in the project had to adhere to strict guidelines and procedures set by NASA. They were required to document every aspect of the design and construction process, including materials used, manufacturing techniques, and quality control measures. This documentation was crucial for ensuring the traceability and accountability of the spacecraft's components.

Quality control measures included non-destructive testing, such as X-ray and ultrasound inspections, to detect any potential flaws or defects in the spacecraft's structure. The contractors also implemented rigorous testing procedures for the spacecraft's systems, including propulsion, life support, and communication.

The collaboration between NASA and the contractors was essential for addressing any issues or challenges that arose during the design and construction process. Regular communication and feedback allowed for continuous improvement and refinement of the spacecraft's design.

The dedication and expertise of the engineers and technicians involved in designing and building the Apollo spacecraft were instrumental in the success of the Apollo 11 mission. Their attention to detail, innovative solutions, and commitment to quality ensured that the spacecraft could withstand the rigors of space travel and safely transport the astronauts to the moon and back.

The next section will explore the rigorous testing and quality control measures implemented to ensure the reliability and safety of the Apollo spacecraft.

11.3 Testing and Quality Control

Before the historic Apollo 11 mission could take place, rigorous testing and quality control measures were implemented to ensure the safety and success of the mission. The engineers and technicians at NASA worked tirelessly to design, build, and test the spacecraft, leaving no room for error.

11.3.1 Testing the Spacecraft

The testing phase of the Apollo 11 mission was a critical step in the process. The spacecraft, consisting of the Command Module, the Lunar Module, and the Saturn V rocket, underwent extensive testing to ensure that every component was functioning properly.

One of the most crucial tests was the Command Module's heat shield, which protected the astronauts during reentry into Earth's atmosphere. The heat shield was subjected to intense heat and pressure simulations to ensure its ability to withstand the extreme conditions of reentry.

The Lunar Module, which would carry Neil Armstrong and Buzz Aldrin to the lunar surface, also underwent rigorous testing. The descent and ascent engines were tested extensively to ensure their reliability and performance in the lunar environment.

Additionally, the Saturn V rocket, the most powerful rocket ever built, underwent numerous tests to ensure its structural integrity and performance. The engines were fired multiple times to simulate the conditions of launch and ascent.

11.3.2 Quality Control Measures

Quality control was of utmost importance in the Apollo 11 mission. Every component of the spacecraft was meticulously inspected and tested to meet the highest standards of safety and reliability.

NASA implemented a comprehensive quality control program that involved inspections, audits, and documentation at every stage of the mission. Each component was subjected to rigorous quality checks to ensure that it met the specified requirements.

The engineers and technicians responsible for building the spacecraft followed strict protocols and procedures to maintain quality control. They adhered to detailed specifications and guidelines to ensure that every component was manufactured to the highest standards.

Furthermore, NASA established a system of checks and balances to verify the quality of the spacecraft. Independent teams of engineers and technicians were assigned to review and validate the work of their colleagues, ensuring that no errors or defects went unnoticed.

11.3.3 Simulations and Mock Missions

To further test the spacecraft and prepare the astronauts for the mission, NASA conducted a series of simulations and mock missions. These exercises allowed the astronauts and mission control to practice various scenarios and identify any potential issues or challenges.

Simulations were conducted in realistic environments that replicated the conditions of space travel. The astronauts practiced docking maneuvers, lunar landings, and emergency procedures in simulators that closely resembled the actual spacecraft.

Mock missions were also conducted to simulate the entire mission from launch to landing. These dress rehearsals allowed the astronauts and mission control to work together and refine their procedures and protocols.

Through these simulations and mock missions, the astronauts gained valuable experience and confidence in their ability to carry out the mission. Mission control also had the opportunity to fine-tune their operations and ensure that they were prepared for any eventuality.

11.3.4 Ensuring Mission Success

The testing and quality control measures implemented for the Apollo 11 mission played a crucial role in ensuring its success. By subjecting the spacecraft to rigorous testing and adhering to strict quality control protocols, NASA minimized the risk of failure and maximized the chances of a safe and successful mission.

The dedication and attention to detail exhibited by the engineers and technicians involved in the testing and quality control processes were instrumental in the overall success of the Apollo 11 mission. Their commitment to excellence and their unwavering pursuit of perfection set the stage for one of humanity's greatest achievements.

As the testing and quality control phase came to an end, the stage was set for the historic Apollo 11 mission. The astronauts, Neil Armstrong, Buzz Aldrin, and Michael Collins, along with the support of mission control and the critical contributions of black women mathematicians, were ready to embark on a journey that would forever change the course of history.

11.4 Supporting the Mission

The success of the Apollo 11 mission relied on the collective efforts of numerous individuals and teams working behind the scenes. From the engineers and technicians who designed and built the spacecraft to the support staff in mission control, each person played a crucial role in supporting the mission and ensuring its success.

11.4.1 The Engineers and Technicians

Behind every successful space mission lies a team of dedicated engineers and technicians who work tirelessly to design and build the spacecraft. For Apollo 11, this team was no exception. Led by the engineers at NASA's Manned Spacecraft Center in Houston, Texas, they were responsible for turning President Kennedy's vision into a reality.

The engineers and technicians faced numerous challenges during the development of the Apollo spacecraft. They had to overcome technical hurdles, such as designing a spacecraft capable of withstanding the harsh conditions of space and developing a reliable propulsion system for the journey to the moon. They also had to ensure the safety of the astronauts by implementing rigorous quality control measures and conducting extensive testing.

11.4.2 Designing and Building the Spacecraft

The design and construction of the Apollo spacecraft was a monumental task that required the collaboration of various teams. The Command Module, which housed the astronauts during their journey to and from the moon, was designed to provide a safe and comfortable environment. It was equipped with life support systems, communication equipment, and navigation instruments to ensure a successful mission.

The Lunar Module, also known as the "Eagle," was responsible for landing the astronauts on the moon's surface. It had to be lightweight yet sturdy enough to withstand the lunar landing and takeoff. The engineers had to carefully consider every aspect of the Lunar Module's design, from its landing gear to its propulsion system, to ensure a successful touchdown and liftoff from the moon.

11.4.3 Testing and Quality Control

Before any spacecraft could be deemed flight-worthy, it had to undergo rigorous testing and quality control procedures. The engineers and technicians conducted countless tests to ensure that every component of the spacecraft functioned as intended.

One of the most critical tests was the Apollo 11 mission simulation, where the spacecraft and its systems were put through a series of simulated missions to mimic the conditions of the actual mission. This allowed the engineers and astronauts to identify any potential issues and make necessary adjustments before the actual launch.

Quality control was of utmost importance throughout the entire process. Every component of the spacecraft was meticulously inspected and tested to ensure its reliability and safety. The engineers and technicians worked tirelessly to meet the stringent standards set by NASA, leaving no room for error.

11.4.4 Supporting the Mission

While the engineers and technicians were responsible for designing and building the spacecraft, there were many other individuals and teams who played a vital role in supporting the mission. From the scientists who analyzed the data collected during the mission to the medical personnel who monitored the astronauts' health, each person had a specific role to play.

Mission control, located at NASA's Manned Spacecraft Center, was the nerve center of the Apollo 11 mission. It was here that flight directors and support staff monitored the spacecraft's systems, communicated with the astronauts, and made critical decisions in real-time. The flight directors, including Gene Kranz and Glynn Lunney, led the teams responsible for ensuring the mission's success.

The support staff in mission control included experts in various fields, such as navigation, communications, and life support systems. They worked together to provide the astronauts with the necessary information and support throughout the mission. Their expertise and quick thinking were crucial in resolving any issues that arose during the mission, ensuring the safety and success of the astronauts.

In addition to mission control, there were numerous other teams involved in supporting the mission. These included the recovery team, responsible for retrieving the astronauts and the spacecraft after splashdown, and the quarantine team, who ensured that the astronauts were not exposed to any potential lunar pathogens upon their return.

The success of the Apollo 11 mission was a testament to the dedication, expertise, and teamwork of all those involved. From the engineers and technicians who designed and built the spacecraft to the support staff in mission control, each person played a vital role in supporting the mission and making history. Their tireless efforts and unwavering commitment ensured that Neil Armstrong, Buzz Aldrin, and Michael Collins could take that one giant leap for mankind and inspire generations to come.

12 The Impact of Apollo 11

12.1 The Global Reaction

The successful Apollo 11 moon landing on July 20, 1969, was a monumental achievement that captivated the world and sparked a global reaction unlike anything seen before. As Neil Armstrong took his historic first step onto the lunar surface, people from all corners of the globe watched in awe and wonder. The impact of this extraordinary feat was felt not only in the United States but also across international borders, leaving an indelible mark on the collective consciousness of humanity.

12.1.1 Awe and Inspiration

The global reaction to the Apollo 11 mission was one of awe and inspiration. People around the world were captivated by the sheer audacity and ambition of the endeavor. The fact that human beings had managed to travel to another celestial body and set foot on its surface was a testament to the power of human ingenuity and determination. The images and footage transmitted back to Earth by the astronauts were beamed into living rooms across the globe, bringing the wonder of space exploration into the homes of millions.

12.1.2 Unity and International Cooperation

The Apollo 11 mission also fostered a sense of unity and international cooperation. Despite the intense competition between the United States and the Soviet Union during the Cold War, the moon landing was seen as a triumph for all of humanity. People from different nations and cultures came together to celebrate this remarkable achievement, transcending political and ideological boundaries. The shared goal of exploring the unknown depths of space brought nations closer together and highlighted the potential for collaboration on a global scale.

12.1.3 Scientific and Technological Advancements

The global reaction to the Apollo 11 mission was not limited to awe and inspiration; it also sparked a renewed interest in scientific and technological advancements. The moon landing demonstrated the incredible capabilities of human beings and pushed the boundaries of what was thought possible. The scientific discoveries made during the mission, such as the collection of lunar samples and the study of the moon's geology, opened up new avenues of research and expanded our understanding of the universe.

12.1.4 Cultural and Artistic Impact

The impact of the Apollo 11 mission extended beyond the realms of science and technology. It had a profound influence on popular culture and the arts. The moon landing inspired countless works of literature, music, and film, becoming a symbol of human achievement and exploration. Artists and writers drew inspiration from the bravery and determination of the astronauts, weaving their stories into the fabric of our cultural heritage. The iconic image of Buzz Aldrin standing on the lunar surface, with the American flag planted beside him, became an enduring symbol of human triumph.

12.1.5 Educational and Inspirational Legacy

Perhaps one of the most significant legacies of the Apollo 11 mission was its impact on education and inspiring future generations. The moon landing sparked a renewed interest in science, technology, engineering, and mathematics (STEM) fields. Children around the world were captivated by the idea of space exploration and aspired to become astronauts, scientists, and engineers. The mission served as a catalyst for educational initiatives and programs aimed at nurturing young minds and encouraging them to pursue careers in STEM.

12.1.6 Political and Geopolitical Implications

The global reaction to the Apollo 11 mission also had political and geopolitical implications. The United States, having achieved a significant victory in the space race, solidified its position as a global superpower. The success of the mission bolstered American prestige and influence on the world stage. It also had an impact on the Cold War rivalry between the United States and the Soviet Union, with the moon landing serving as a symbolic victory for the United States in the race for technological and scientific superiority.

12.1.7 The Human Spirit and Exploration

Above all, the global reaction to the Apollo 11 mission highlighted the indomitable spirit of human exploration. It demonstrated that when faced with seemingly insurmountable challenges, humanity has the capacity to overcome them through innovation, collaboration, and sheer determination. The moon landing was a testament to the human spirit's insatiable curiosity and desire to push the boundaries of what is known and explore the unknown.

The impact of the Apollo 11 mission was felt far beyond the borders of the United States. It united people from all walks of life and inspired generations to dream big and reach for the stars. The global reaction to this historic event serves as a reminder of the power of human achievement and the limitless possibilities that lie within our grasp.

12.2 Space Exploration and International Cooperation

Space exploration has always been a global endeavor, and the Apollo 11 mission was no exception. The historic moon landing of July 20, 1969, not only marked a significant achievement for the United States but also represented a triumph of international cooperation and collaboration.

12.2.1 The Cold War Context

At the time of the Apollo 11 mission, the world was in the midst of the Cold War, a period of intense rivalry between the United States and the Soviet Union. The space race between these two superpowers was a key aspect of this competition, with each nation striving to demonstrate its technological and scientific prowess.

12.2.2 The United States and the Soviet Union

While the United States and the Soviet Union were engaged in a fierce competition to be the first to reach the moon, there were also moments of cooperation and collaboration between the two nations. In 1967, the United States and the Soviet Union signed the Outer Space Treaty, which aimed to ensure the peaceful exploration and use of outer space. This treaty laid the foundation for future international cooperation in space exploration.

12.2.3 International Tracking Network

One of the most significant examples of international cooperation during the Apollo 11 mission was the establishment of the International Tracking Network. This network consisted of tracking stations located around the world, operated by various countries including Australia, Spain, and France. These stations played a crucial role in tracking the Apollo spacecraft and providing vital communication links between mission control and the astronauts.

12.2.4 Lunar Receiving Laboratory

Following the successful moon landing, the United States opened the Lunar Receiving Laboratory (LRL) at the Manned Spacecraft Center in Houston, Texas. The LRL was responsible for the safe handling and analysis of the lunar samples brought back by the Apollo missions. Scientists from around the world were invited to collaborate on the analysis of these samples, further promoting international cooperation in lunar research.

12.2.5 Apollo-Soyuz Test Project

In 1972, the United States and the Soviet Union embarked on a groundbreaking joint mission known as the Apollo-Soyuz Test Project (ASTP). This mission involved the docking of an American Apollo spacecraft with a Soviet Soyuz spacecraft in Earth orbit. The ASTP represented a significant milestone in space exploration as it marked the first time that American and Soviet astronauts worked together in space.

12.2.6 Scientific Collaboration

The Apollo missions not only aimed to put humans on the moon but also had a strong focus on scientific exploration. The lunar samples brought back by the Apollo astronauts provided scientists from around the world with invaluable insights into the moon's geology and history. Researchers from various countries collaborated on the analysis of these samples, contributing to our understanding of the moon and its formation.

12.2.7 Cultural Exchange

The Apollo 11 mission also had a profound impact on cultural exchange between nations. The images and videos of the moon landing were broadcast worldwide, captivating audiences across the globe. The achievement of landing humans on the moon inspired people from all walks of life and transcended national boundaries. It served as a reminder of the shared aspirations and dreams of humanity.

12.2.8 Legacy of International Cooperation

The international cooperation and collaboration witnessed during the Apollo 11 mission laid the foundation for future space exploration endeavors. It demonstrated that even in the midst of political tensions, nations could come together to achieve common goals and push the boundaries of human exploration. The spirit of cooperation fostered during the Apollo era continues to shape international space missions to this day.

In conclusion, the Apollo 11 mission not only represented a significant milestone in human history but also showcased the power of international cooperation. Despite the Cold War context, the United States and the Soviet Union found common ground in their pursuit of space exploration. The establishment of the International Tracking Network, the collaboration in lunar sample analysis, and the Apollo-Soyuz Test Project all exemplify the spirit of cooperation that defined this era. The legacy of international cooperation continues to inspire and guide future space exploration missions, reminding us of the potential for unity and collaboration in the pursuit of scientific discovery and human achievement.

12.3 The Space Race Comes to an End

The Space Race, which began with the launch of the Soviet satellite Sputnik in 1957, reached its climax with the historic Apollo 11 moon landing in 1969. This monumental achievement marked the end of the intense competition between the United States and the Soviet Union to demonstrate their technological and ideological superiority. The success of Apollo 11 not only fulfilled President Kennedy's vision of landing a man on the moon but also brought about a significant shift in the dynamics of the Space Race.

12.3.1 The Changing Landscape

By the late 1960s, the United States had made significant strides in space exploration. The Mercury and Gemini programs had paved the way for the ambitious Apollo program, which aimed to land astronauts on the moon. The Soviet Union, on the other hand, had achieved several milestones in space, including the first manned spaceflight by Yuri Gagarin in 1961. However, they faced setbacks in their lunar exploration efforts, with multiple failed attempts to land a spacecraft on the moon.

The Apollo 11 mission represented the culmination of years of scientific research, technological advancements, and meticulous planning. The successful landing of Neil Armstrong and Buzz Aldrin on the lunar surface on July 20, 1969, was a momentous achievement for the United States and a significant blow to the Soviet Union's space ambitions.

12.3.2 A New Era of Cooperation

While the Space Race had been characterized by fierce competition and a race for supremacy, the success of Apollo 11 opened up new possibilities for international cooperation in space exploration. The United States extended an invitation to the Soviet Union to collaborate on a joint lunar mission, but the offer was declined. Nevertheless, the spirit of cooperation began to emerge in subsequent years.

In 1972, the United States and the Soviet Union signed the Apollo-Soyuz Test Project, a groundbreaking mission that saw an American Apollo spacecraft dock with a Soviet Soyuz spacecraft in orbit. This historic mission marked the first time that American and Soviet astronauts worked together in space, symbolizing a thawing of tensions and a shift towards collaboration.

12.3.3 The Decline of the Soviet Space Program

The success of Apollo 11 dealt a significant blow to the Soviet Union's space program. The Soviet Union had been the first to achieve several milestones in space exploration, including the first manned spaceflight and the first woman in space. However, their failures in lunar exploration and the United States' successful moon landing highlighted the technological gap between the two nations.

In the years following Apollo 11, the Soviet Union faced a series of setbacks and budget constraints that hampered their space program. The focus shifted from manned lunar missions to unmanned missions and space station development. While the Soviet Union continued to make significant contributions to space exploration, including the launch of the first space station, Mir, in 1986, they were no longer at the forefront of the space race.

12.3.4 The Enduring Legacy

The legacy of Apollo 11 extends far beyond the end of the Space Race. The successful moon landing not only demonstrated the technological prowess of the United States but also inspired generations of scientists, engineers, and dreamers around the world. The images and footage of Neil Armstrong taking his first steps on the lunar surface captured the imagination of people everywhere and left an indelible mark on human history.

The Apollo program also had a profound impact on scientific research and technological advancements. The mission brought back valuable lunar samples, providing scientists with a wealth of information about the moon's geology and history. The technological innovations developed for the Apollo program, such as lightweight materials, miniaturized electronics, and advanced guidance systems, found applications in various industries and continue to shape our lives today.

Furthermore, the Apollo 11 mission highlighted the critical contributions of black women mathematicians, such as Katherine Johnson, Dorothy Vaughan, and Mary Jackson, who played pivotal roles in the success of the Apollo program. Their remarkable achievements, despite facing racial and gender discrimination, shattered barriers and paved the way for future generations of women and minorities in STEM fields.

In conclusion, the success of the Apollo 11 moon landing marked the end of the Space Race and ushered in a new era of international cooperation in space exploration. The United States' achievement not only solidified its position as a leader in space but also inspired the world with the possibilities of human exploration beyond Earth. The enduring legacy of Apollo 11 continues to shape our understanding of the universe and serves as a testament to the power of human ingenuity and determination.

12.4 Apollo 11's Enduring Legacy

The Apollo 11 mission, with its successful moon landing on July 20, 1969, left an indelible mark on human history. The enduring legacy of this monumental achievement can be seen in various aspects of society, from scientific advancements to technological breakthroughs and the inspiration it provided to future generations.

12.4.1 Scientific Discoveries and Lunar Samples

One of the most significant legacies of Apollo 11 is the scientific knowledge gained from the mission. The astronauts collected valuable lunar samples during their time on the moon, bringing back approximately 47 pounds of rocks and soil. These samples provided scientists with a wealth of information about the moon's composition, geology, and history.

Through extensive analysis, researchers discovered that the moon's surface is primarily made up of basalt, a type of volcanic rock. This finding helped scientists understand the moon's volcanic past and its geological evolution. Additionally, the lunar samples provided evidence of the moon's age, which is estimated to be around 4.5 billion years, similar to the age of Earth.

The analysis of the lunar samples also revealed the presence of water molecules, albeit in small quantities. This discovery challenged the previous belief that the moon was completely devoid of water. It opened up new possibilities for future lunar exploration and the potential for utilizing lunar resources.

12.4.2 Technological Advancements

The Apollo 11 mission pushed the boundaries of technology and engineering, leading to numerous advancements that have had a lasting impact. The development of the Saturn V rocket, the command module, and the lunar module showcased the incredible capabilities of human ingenuity and innovation.

The Saturn V rocket, standing at a towering height of 363 feet, remains the most powerful rocket ever built. Its successful launch and subsequent missions paved the way for future space exploration endeavors. The command module, named Columbia, served as the astronauts' living quarters during their journey to the moon and back. Its design and functionality set the standard for subsequent spacecraft.

The lunar module, known as the Eagle, played a crucial role in the moon landing. Its ability to separate from the command module, descend to the lunar surface, and then ascend to dock with Columbia demonstrated the precision and reliability of the engineering involved. The technology developed for the lunar module laid the foundation for future lunar landers and contributed to the advancement of space exploration.

Furthermore, the Apollo Guidance Computer (AGC), a groundbreaking computer system developed for the Apollo missions, was a significant technological achievement. It was the first computer to use integrated circuits, making it smaller, lighter, and more powerful than previous systems. The AGC's capabilities and reliability were instrumental in the success of the Apollo 11 mission and set the stage for the development of modern computer technology.

12.4.3 Inspiring Future Generations

The Apollo 11 mission captured the imagination of people around the world and inspired a new generation of scientists, engineers, and explorers. The sight of Neil Armstrong taking his first steps on the lunar surface was a testament to human determination and the limitless possibilities of human achievement.

The success of Apollo 11 demonstrated that seemingly insurmountable challenges could be overcome through dedication, teamwork, and innovation.

It ignited a sense of wonder and curiosity in the minds of young people, encouraging them to pursue careers in science, technology, engineering, and mathematics (STEM).

The mission also highlighted the importance of international collaboration and cooperation. The space race between the United States and the Soviet Union had been marked by competition and rivalry. However, the Apollo 11 mission showcased the potential for nations to come together and achieve extraordinary feats for the benefit of all humanity.

12.4.4 The Lasting Impact

The legacy of Apollo 11 extends far beyond the initial moon landing. It laid the foundation for future space exploration missions and paved the way for the establishment of the International Space Station (ISS) and other collaborative ventures.

The technological advancements made during the Apollo program have found applications in various fields, including telecommunications, satellite technology, and medical imaging. The miniaturization of electronics, advancements in materials science, and the development of life support systems have all contributed to advancements in everyday life.

Moreover, the Apollo 11 mission served as a symbol of human achievement and the triumph of the human spirit. It reminded us of the power of setting audacious goals and working tirelessly to achieve them. The mission continues to inspire individuals and nations to reach for the stars and explore the unknown.

In conclusion, the enduring legacy of Apollo 11 is multifaceted and far-reaching. From the scientific discoveries and technological advancements to the inspiration it provided to future generations, the mission's impact on society cannot be overstated. The moon landing was a testament to human ingenuity, perseverance, and the unyielding desire to explore the unknown.

The legacy of Apollo 11 will continue to shape the future of space exploration and serve as a reminder of what can be achieved when humanity dares to dream big.

www.ingramcontent.com/pod-product-compliance
Lightning Source LLC
Chambersburg PA
CBHW072159290526
45794CB00004B/1571